The Fragile Edge

BOOKS BY JULIA WHITTY

A Tortoise for the Queen of Tonga

The Fragile Edge

Julia Whitty

The Fragile Edge

Diving and Other Adventures
in the South Pacific

Houghton Mifflin Company

BOSTON   NEW YORK   2007

For information about permission to reproduce selections
from this book, write to Permissions, Houghton Mifflin Company,
215 Park Avenue South, New York, New York 10003.

Visit our Web site: www.houghtonmifflinbooks.com.

Library of Congress Cataloging-in-Publication Data
Whitty, Julia.
    The fragile edge : diving and other adventures in the South Pacific /
Julia Whitty.
        p.   cm.
    ISBN-13: 978-0-618-19716-3
    ISBN-10: 0-618-19716-8
    1. Coral reef ecology — French Polynesia — Rangiroa. 2. Rangiroa
(French Polynesia) I. Title.
QH198.F74W48 2007
578.77'89099622 — dc22   2006023706

Book design by Melissa Lotfy

Maps by Sharon Urquhart

Printed in the United States of America

MP 10 9 8 7 6 5 4 3 2 1

Portions of this book first appeared as "Shoals of Time," in *Harper's Magazine,*
January 2001; "All the Disappearing Islands," in *Mother Jones,* July / August 2003;
and "The Fate of the Ocean," in *Mother Jones,* March / April 2006.

Epigraph from an untitled poem by Rumi from *The Essential Rumi,* Castle
Press, translated by Coleman Barks. Copyright © Coleman Barks. Reprinted
by permission of the author.

Excerpts from *The Ink Dark Moon: Love Poems by Ono no Komachi and Izumi Shikibu, Women of the Ancient Court of Japan,* copyright © 1990 by Jane Hirshfield,
translator. Used by permission of Random House, Inc.

Lines from the poem by Sami Manzei, translated by Edwin A. Cranston, from
A Waka Anthology: Volume One: The Gem Glistening Cup, copyright © 1993 by the
Board of Trustees of the Leland Stanford Jr. University. All rights reserved.
Used with the permission of Stanford University Press.

For Ian
   with love and wonder

ACKNOWLEDGMENTS

Many thanks to many, including to Hardy Jones, who led my way into the underwater world, and who appears, namelessly, throughout these pages; to Clara Jeffrey, who opened the doors of the literary world to me; to Patience Whitty and Tim Whitty for lifelong encouragement; to John Whitty for ongoing translations from the perplexing language of mathematics; to Lillian Howan and Bruno Leou-on for graciously introducing me to Tahiti, and for their help with French and Tahitian translations, to Mary Mahood for braving the first reading; with all its attendant hazards, to Melanie Jackson for her good agenting and good advice; to Deanne Urmy for sensitive editing in the breach; to Reem Abu-Libdeh for delicate copyediting; and to the late Rona Jaffe and the Rona Jaffe Foundation for generous support. Infinite thanks to Ian Riedel for his bright and lasting faith in me.

Contents

Part I: Rangiroa

1. Rapture ~ 3
2. Swimming in the Bellybutton ~ 11
3. The Manysided Lagoon ~ 16
4. Changeover ~ 22
5. Breath Control ~ 28
6. Inside the Turtle's Shell ~ 36
7. The Near-Field/Far-Field Boundary ~ 44
8. Eavesdropping ~ 52
9. Big Songs ~ 58
10. The King of Lake Vaihiria ~ 65
11. Impermanence ~ 74
12. The Infinity Pool ~ 81
13. Inshallah ~ 89
14. Poi Dogs ~ 94
15. The Consorting Together of Dissimilar Organisms ~ 103
16. It Furthers One to Cross the Great Water ~ 108
17. The Lemon Shark Affair ~ 114
18. Grand Secret ~ 121

Part II. Funafuti

19. Hideaway ~ 133
20. Falling Dominoes ~ 140
21. Liquid Faultline ~ 147
22. Leave Your Values at the Front Desk ~ 153
23. Little Cemeteries ~ 158
24. Diving the Apocalypse ~ 166
25. Nuptials ~ 173
26. Just Do It ~ 181
27. Beseeching the Wind Horses ~ 187
28. Sinking Dragons ~ 193

Part III. Moʻorea

29. The Churning of the Ocean ~ 201
30. An Ocean of Silence and Bliss ~ 208
31. The Sleep of Plants ~ 214
32. Living Lanterns ~ 219
33. The Clamor of True Democracy ~ 224
34. The Spirit of Godly Gamesomeness ~ 231
35. Coral Noose ~ 238
36. Mother Ocean ~ 246
37. Fish Tamer ~ 254
38. A Force Like a Hundred Thousand Wedges ~ 259
39. Gleanings ~ 267
40. Across the Threshold ~ 272

Epilogue ~ 277
Notes ~ 282
Glossary ~ 287

Late, by myself, in the boat of myself,
no light and no land anywhere,
cloudcover thick. I try to stay
just above the surface, yet I'm already under
and living within the ocean.

— *Mawlana Jalal ad-Din Muhammad Rumi,*
   1207–1273

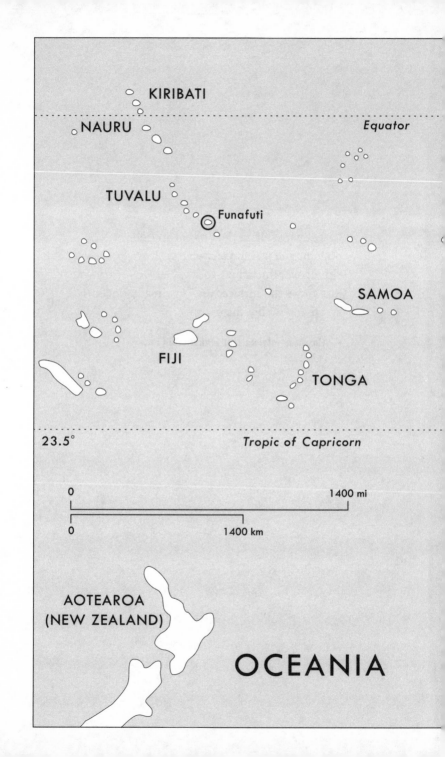

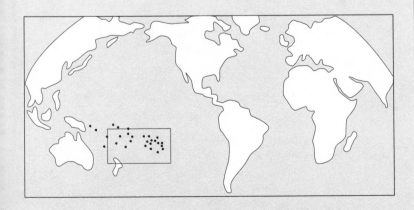

0°

Pacific Ocean

Marquesas Islands

Tuamotu Archipelago

Rangiroa

Teti'aroa

Mo'orea

Tahiti

FRENCH POLYNESIA

Society Islands

Moruroa

Austral Islands

Gambier Islands

Part I

Rangiroa
Tuamotu Archipelago

We feel surprise when travellers tell us of the vast dimensions of the Pyramids and other great ruins, but how utterly insignificant are the greatest of these, when compared to these mountains of stone accumulated by the agency of various minute and tender animals! This is a wonder which does not at first strike the eye or the body, but, after reflection, the eye of reason.

— *Charles Darwin upon seeing the volcano-and-coral constructed isle of Mauritius, 1836*

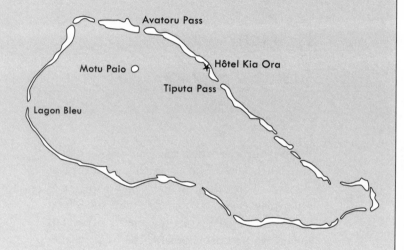

Avatoru Pass

Motu Paio ○

Hôtel Kia Ora

Tiputa Pass

Lagon Bleu

0   10
Approximate Miles

RANGIROA

1

Rapture

ALL DAY WE HAVE been observing the surgeonfish, which the Tahitians call *maroa* and the French call *chirurgien,* streaming over the outer reef slopes in tightly knit single files. Now, in the final hour of daylight, scores of them coalesce into banners of yellow-and-blue fins, flowing gaudily over the contours of the coral toward the edge of Tiputa Pass. Bunched into agitated crowds of hundreds, they rise above the reefs, swarming and butting each other in a chaos of seeming cross purposes.

Stimulated by the onset of a waning moon in the South Pacific, the surgeonfish cluster, rise, bump, then drop back to the reef, disperse, circle, regroup, and rise again. A dozen times they practice, each round taking them higher into the water column, farther from the safety of the coral. The foreplay culminates in what scientists call spawning and what the French divers I'm with charmingly refer to as lovemaking — a pair of surgeonfish detaching from the crowd and exploding upward in an impossibly fast arc, then ejecting their sperm and eggs into the open water in a burst of milky smoke. Never breaking stride, the pair shoots back to the reef at speeds nearly unrecordable by the human eye. Other pairs follow. And others. At the apex of each upward burst, the ejaculated white puff-balls hang still, yet riotously mobilized as the chemistry of conception begins, sperm seeking eggs with only a moment for the microjourney to succeed before the gametes are caught in the outflow of water from the pass, torn apart, and carried out to deep water.

When it's too dark underwater to see anymore, we motor back to land against a sunset as soft and mutable as a watercolor. While we rinse our gear at the dock, my fellow divers have blissful expressions on their faces. It's always this way. It doesn't matter who they are, what their names are, or where the coral reef is; but in this case, as we return from Tiputa Pass at Rangiroa Atoll in French Polynesia, the faces are those of two young Frenchmen. Both guide here professionally, at this fragile edge between land and sea. If anyone would be jaded it would rightfully be them. But the opposite has happened. The rapturous state induced by the reef has caught these tough young men in its spell and rendered them, for this moment, as beatific as angels.

It's rare in the wild to see the moment of conception, and our mood is joyful, relieved, as if we've conquered some summit and survived. The French divers are happy this day delivered up such a pretty secret. Back in my *fare* (bungalow), I check my underwater slate, eager to transcribe its notes into my journal — only to laugh when I discover nothing more than a single exclamation point marked on it.

Such is the paradox of the reef: a world that feels purely and extravagantly sensual yet exists mostly outside our own sensory realm. We smell nothing underwater (although the sea is filled with scents), taste only the metallic twang of compressed air, see poorly, and are reduced to nondirectional hearing; in effect, we're disabled. Nor can we talk. Without language, without the correct words, or any words, human divers revert to a preverbal state of mind. So the dive you've just made tends to be felt rather than accurately remembered, and the little plastic slate you've dutifully carried underwater for notetaking reveals only doodles or strange hieroglyphs — made more difficult to decipher since the pencil marks, supposedly erasable by scrubbing with beach sand, never really do come clean, leaving you to contend with the ghostly outlines of all your previous, equally enigmatic dives.

This is the struggle, or at least my struggle, working underwater: how to comprehend and then translate the otherworldly marvels under the surface into the alien world topside.

On all of planet earth exist only three hundred thirty coral atolls (Malayam *atolu:* reef; from *adal:* closing, uniting), the necklace-shaped islands consisting of sandy islets surrounding a tropical lagoon. Rangiroa is one of the seventy-seven atolls comprising the Tuamotu Archipelago, itself one of the five archipelagos of French Polynesia. Taken together, all the islands of French Polynesia cover an area of the South Pacific larger than western Europe, and the Tuamotus inhabit a seascape larger than California.

Millions of years ago, the Tuamotu Archipelago was the scene of rampant vulcanism fueled by a rising plume of magma known as a hot spot under the Pacific plate. Today these islands sit atop a wracked area of sea floor known as the Tuamotu Ridge, which marks the eastern edge of the volcano-scarred western Pacific. To the east, as far as the Americas, lies an entirely different bathymetry, a relatively featureless underwater plain scored by a series of east-west running fracture zones. The Tuamotus lie between the Marquesas Fracture Zone to the north and the Easter Fracture Zone to the south — the names indicative of some of the Tuamotus' nearest neighbors.

This is an isolated place. The nearest continental land to Rangiroa Atoll lies four thousand miles due south at Marie Byrd Land in Antarctica, and the nearest inhabited land of any size lies forty-four hundred miles due west at the Cape York Peninsula on Australia's northernmost tip. To find anything substantial in the opposite direction, you have to travel five thousand miles east to Peru. There is hardly a place on the globe farther from continental landmasses, and as a result both the underwater and the topside worlds are dominated by the Pacific Ocean.

The superlatives surrounding this far from peaceful body of water are daunting. At sixty-four million square miles, it's larger than the total land area of the world combined, covers a full one-third of our planet, and contains more than half the world's seawater. It is home to thirty thousand islands and five climate regions — the westerlies, the trades, the monsoon region, the typhoon region, and the doldrums. Its basin is deeper than that of any other ocean, maintaining an average draft of fourteen thousand feet, and in places plunging abysmally to a world-record-setting thirty-six thousand feet, about seven miles. The Pacific encompasses the epicenter of coral evolution today, and is home to 40 percent of all the world's reefs, including our planet's most extensive barrier reef and atoll formations. On this unfathomably productive and dangerous seascape, tiny Rangiroa floats like a fallen flower petal — seemingly too small to survive or to be survived upon.

Yet this is the second largest atoll on earth, stretching roughly fifty miles from east to west and twenty miles from north to south. It encloses a nearly four-hundred-square-mile lagoon within a one-hundred-forty-mile bracelet of islets — four hundred eighteen islets in all. Called *motu* by the Tahitians, these islets are almost as insignificant as midoceanic shoals, since they rise only fifteen feet above sea level, are virtually devoid of freshwater, and possess only sand for soil. Yet they have been tenaciously colonized by Polynesian people throughout the South Pacific. In fact, the sandy *motu* of atoll islands have always welcomed open-ocean sailors, including the drift-seeds and drift-fruits that bob on the waves: the coconut, the mangroves, and their unwitting passengers of insects, crabs, and seabirds.

To stare out at Rangiroa's lagoon is akin to looking over an infinite ocean, with the *motu* on the far side lost beneath the curvature of the earth. In fact, the only clue you have that it is not the ocean is its relative calm. Three hundred feet away on the open-ocean side of the *motu*, the surf thunders day and night without pause, whereas most days the lagoon side is as calm as a millpond, its tranquil waters painted in so many vibrant shades of blue and green and turquoise and beryl that you conclude the ancient Tahitians had it right, that God is not in the clouds but in the lagoon.

Because of its geography, Rangiroa has a tremendously dynamic reef system. Its four hundred eighteen *motu* are separated by tidal channels called *hoa* — most of which are too shallow for anything but the smallest boats to travel through, and some are too small for that. Yet Rangiroa also possesses the crown jewel of atolls — two navigable *hoa*, five hundred yards wide, known as Tiputa Pass and Avatoru Pass.

Partly because of these passes, the changing of the tides at Rangiroa Atoll produces an extraordinary phenomenon of pressures and counterpressures as seawater either presses into the lagoon or rapidly escapes from it. At these times, Tiputa Pass in particular is transformed into a maelstrom of inflow and outflow, water colliding and spinning off whirlpools that appear underwater as snaky tornadoes of blinding sediment. The French call this the *mascaret*, translated as a high wave that travels backward up large rivers during flood tides. In Rangiroa, the word refers to conditions that occur when the tide is flowing out of the lagoon into the open sea and the wind is blowing against it, or vice versa. This happens once or twice a day when the battle between wind-driven currents and water-driven currents result in waves climbing each other's faces and battling to a standstill — in other words, becoming standing waves, which in Tiputa Pass can reach fifteen feet.

No human diver can survive immersion into the full brunt of the *mascaret*. So those of us who come to Rangiroa dive its edges, where the noise of the rip tide rumbles beneath the brighter soundtrack of the sea, the snapping, clicking, rasping, buzzing, squealing, and grunting of fish, shrimp, clams, and corals at work. For these creatures, the waxing and waning of the *mascaret* is an external force akin to a combined circulatory and respiratory system — an enormous set of oceanic lungs inhaling and exhaling four times a day, pausing briefly in the slacktide. In this way the sea is enriched: huge loads of organic matter emptying out of the lagoon to nourish the outer reef slopes, immeasurable quantities of clear water from the open sea rushing back to refresh it.

As with many of my dives in the course of shooting four nature documentaries here, Yann Hubert is guiding the way. I'm familiar with his yogilike presence underwater. Simply put, he's the best diver I've ever seen — a total contrast to his dry-world persona, which tends to fidgety boredom, chain-smoking, and amusing himself with incomprehensibly profane and colloquial French jokes. But down here he's a master, breathing so rarely, so delicately, that fish sidle up to him as if he were not, like the rest of us, an intruder from another world. I believe I understand his secret: total compliance with the environment. Rather than challenge the currents he remains still, using the minimum angulation of body or fins to move. But even knowing this, I'm stunned to see — after surreptitiously checking his gauges at the end of several dives — that he doesn't consume enough air to sustain human life.

This morning, we descend through a crowd of yellowspot emperors resting in the shade cast by our inflatable Zodiac boat, then beyond them, through a phalanx of batfish, steely and triangular as arrowheads. Yann drifts to my left, falling lightly through the center of a silvery cyclone of thousands of blackfin barracuda riding one another's tails in a vortex of their own making. They open up to let him through, then close ranks behind him.

Far below the coral appears, mottling the sandy bottom in deepening shades of blue. Large animals drift over it; a manta ray sailing the edge of the current, ghostly schools of gray reef sharks, hundreds packed shoulder to shoulder, shoaling like lowly sardines. This is their chief defense against even larger predators, and proof that there is no absolute invincibility in the sea. A few rise to greet us, their blunt bodies cruising through schools of bluelined snappers.

This is the season when great hammerhead sharks (*Sphyrna mokarran*), which the French call *grand requin marteau* and the Tahitians call *mao tuamata,* cruise Tiputa Pass, snacking on reef sharks exhausted after bloody mating rituals. Although the two cameramen I'm with are busy at the moment filming the cyclone of barracuda and a large, friendly napoléon wrasse, we are all hoping that one of the ten- or twelve-foot (or more) hammerheads will sidewind its way over the bottom and frighten the reef sharks into acting like a school of oversized minnows.

Yet there are many other creatures of interest, including a small squadron of spotted eagle rays sailing the wall of coral below. These are among the most graceful animals in the sea, adorned with Art Deco designs of white circles and polka dots on velvety black backgrounds. They appear to be all wings and soar as hydrodynamically as underwater kites. This group is in nuptial flight, with seven males in pursuit of a female.

I wave at the cameramen, one below me, one above. But they are focused on their own tasks and blinded by their dive masks. I shout, and a profusion of bubbles rises from the regulator in my mouth, drawing a curtain over the scene. Though my voice is deafening in my own head, it cannot travel the short distance to either cameraman through the dense medium of the water. Frustrated, I swim through my bubbles to the downhill cameraman, grab him, and point. He flinches from the contact, as we all do when touched underwater, then follows my finger. Less than a minute has elapsed since I saw the spotted eagle rays, yet already they have faded to faint shadows silkscreened on the distant ocean. He looks in my eyes and shrugs. The rays are barely twitching their wingtips, yet we could never, in our wildest dreams, hope to catch them.

This is only one of what amounts to thousands of missed opportunities underwater — opportunities missed so regularly and predictably that filming or working here becomes a Zen exercise in detachment. If we were to want the eagle rays too much, or to find ourselves filled with regret at having missed them, then our experiences underwater would quickly become bitter ones. Just to test our knowledge of this, we also, on this dive, miss filming a triggerfish building a nest, a traveling circus of bumphead parrotfish, and, on the smallest scale, a starfish shrimp from the genus *Periclimenes* living nearly perfectly camouflaged atop its host sea star.

2
Swimming in
the Bellybutton

MOST OF OUR TIME diving Tiputa Pass we hug the bottom, with its buzzcut of tight, low coral formations, the only kind that can weather the extreme currents here. One of the effects of the *mascaret* is to telescope habitats inside the pass so the creatures of the deep mingle alongside the life of the shallows, making it difficult to know where to look — to focus out to the depths where the big animals dwell, or in to the reef where a metropolis of small life clambers through the coral. Occasionally, something demands your attention in the simplest way, by appearing suddenly in front of you from the blindspots of your peripheral vision. A bluespotted cornetfish (*Fistularia commersonii*) drifts by, as long and narrow as a section of PVC pipe, staggeringly unfishlike. It travels in the manner of its species, vertically, head down, large eyes circling around at me, not alarmed but likewise interested. We study each other.

Suddenly it blushes. Bright aquamarine stains its white head, purple blossoms into its body, as it drifts to a stop alongside an identically colored backdrop: purple coral head bedecked with a flock of blue-green chromis fish. There it hovers, confident in its masquerade, awaiting the opportunity to dart in, and with a jerk of its tubular body, vacuum up the one chromis too slow, foolish, or forgetful to survive this day. I lie on the bottom, awaiting the outcome, rolling gently back and forth in the surge from the distant surface . . . until Yann, silent in the near absence of his scuba bubbles, startles me with a touch on my arm and points to a stonefish the size of a football only a few inches from my knee. It takes me a moment to see — a strange, nearly invisible creature appearing exactly like an algae-encrusted piece of coral rubble. Yet this fish carries enough venom in its dorsal spines to kill me.

Everyone is prey to the disguises of the reef, and Yann picks up a real piece of coral rubble and places it gently on top of the stonefish, assuring himself that I will not unwittingly roll on top of it. Then he points to where the cornetfish is drifting off, the tail of a chromis wriggling in its mouth.

This miraculous place, this atoll crammed with the tense energies of life and death, is the end product of countless trillions of individual coral animals, most no bigger than ants, living and feeding, reproducing and dying, over the course of millions of years. Coral reefs occupy a tiny footprint on the earth: only two hundred thirty thousand square miles, or one-tenth of one percent of our planet's surface. Yet this fragile edge is a wellspring of biodiversity, home to at least four thousand, or nearly a quarter, of all marine fish species. Scientists estimate that coral reefs harbor at least one million species of plants and animals — though some believe this is only the table of contents of a catalog that may well exceed nine million species. "And there are many souls embodied in water," says an ancient Jain text. "Truly water is alive."

Nowhere is this better seen than in the size of coral architecture. Reefs are built by living things, and their architecture, like ours, in its various stages of upkeep or decay, reveals a long archaeological record of life, death, steadfastness, and upheaval. Altered by climate and sea level changes, cyclones, tsunamis, earthquakes, and volcanoes, reefs are the record keepers of the sea. If you compare them to our most ancient things — say Giza, in Egypt, alternately buried and revealed by the shifting grains of the Sahara — so corals slowly build and then bury their own monumental works, along with generations of their dead. Measurements in some Pacific atolls show sediments of ancient reefs stretching nearly a mile deep below the living coral. Yet before each veneer of life succumbs, it thrives as brightly as a souk, complete with laborers, shirkers, hawkers, criminals, tourists, money-lenders, and caliphs.

Tiputa Pass is itself an archaeological relic, offering evidence of many things no longer on this earth, including a long-extinct volcano or volcanoes, a beautiful mountainous island, and a pair of once-mighty rivers. Millions of years ago, a high island like the island of Hawaii stood where today the still waters of Rangiroa's lagoon lap against sandy *motu*. Built by active magma vents beneath the earth's crust, the shield (from the Icelandic *Skjaldbreiður,* "warrior's shield") volcano or volcanoes that formed the ancient Rangiroa Island would have risen twelve thousand feet from the sea floor to the surface, and then another ten thousand or twelve thousand or fifteen thousand feet above sea level. There the summit (or summits) trapped the windy moisture from the surrounding ocean, transforming its energy into afternoon thunderclouds, which in turn fed rivers that roared down the mountain flanks to water forests verdant with native *tiare* and pandanus trees.

We know that at least two such rivers existed on the now non-existent Rangiroa Island because they carved their way down the slopes of the high island, through its barrier reef, and out into the open sea — in the process carrying enough freshwater to prevent any coral growth in sections of the barrier reef that correspond to present-day Tiputa and Avatoru passes. The ghosts of those rivers flow in the form of the *mascaret,* which nowadays runs salty enough for corals to grow, but too swiftly for them to gain a thick toehold, keeping Tiputa and Avatoru passes open.

Or at least they remain open for the time being. Because nothing in Rangiroa, or in the coral world, for that matter, ever remains the same, a fact alluded to on older sailing charts, which refer to the Tuamotus by a variety of ominous-sounding names, including the Dangerous Archipelago, the Rough Waters, and the Labyrinth. Past generations of Western mariners considered these islands too low-lying, and the rate of coral growth too rapid, to expect anything but treachery and heartbreak here. In contrast, the Polynesians, with their sailing/rowing/surf-riding boats, perceived this archipelago as a friendlier place and called it first the Paumotus, the Submissive Islands or Low Islands, and only later — perhaps as the center of power took hold in faraway Tahiti — the Tuamotus, or Distant Islands.

Most days Rangiroa's *mascaret* runs with a certain predictability — although never a total one, I notice, as none of the locals here, whether French or Tahitian, can ever really say what's going to happen, or even what's happening at the moment, in terms of the *mascaret* and all its feeder currents. Sometimes they pretend they can, then shrug it off when they can't (if they're French), or laugh it off good-naturedly (if they're Tahitian).

From a diver's point of view, wrong predictions are a bad thing. Here in Rangiroa, you're always diving currents, and your boatman expects you to emerge from your dive forty-five minutes from now somewhere *over there*. If you get the currents wrong and emerge forty-five minutes later *way-the-other-way-out-there*, it might be that your driver cannot see you and you are being carried faster and harder than a human can swim away from all known land. Or, equally as bad, you may be swept up the backs of the twenty-foot combers breaking on the windward side of the atoll. There is no feeling quite like coming up from your dive to find no boat in sight.

By misdiagnosing the currents at Rangiroa you also run the risk of getting swept up into the *mascaret* itself. This is a frightening zone akin to the extreme whitewater rapids on a river, complete with holes, whirlpools, downcurrents, rip tides, and all the other powerful forces of moving water against which a human is powerless. A few years ago on Fakarava — another Tuamotuan atoll with an even bigger and more dangerous pass — the French owner of the pioneering scuba-diving operation there was killed in the *mascaret*. His body was recovered a few days later by French Navy divers who found it rolling around on the bottom at three hundred feet. No one knows exactly what killed him, but it was probably some lethal intersection between misjudged currents and unimaginably tortured water. To my knowledge, no one has dived that pass since.

Yet as swift and scouring as Rangiroa's *mascaret* runs, it also carries within its storm waters the most delicate seeds of new life in the form of larval fish, larval crustaceans, larval cnidarians, larval mollusks, and all the mostly transparent little things that make up the plankton (Greek *plancto:* wandering) realm. Truly water is alive. Delivered home from their germinal travels in the pelagic zone (Greek *pelagos:* open sea), these plankters come to Rangiroa to give up the nomadic life, settle down, mature, and promulgate their own kind through elaborate and often bizarre reproductive rituals. Settling to the bottom with gluelike anchors, or hiding within the shelter of coral crannies or caves, these juveniles unknowingly add their own biolayer to the layers that have obscured the remains of the high island for millions of years.

Because it's still out there, under the waters of the lagoon — that high island that would once have been a monumental landmark in the watery reaches of the South Pacific. As soon as Rangiroa's volcano went extinct, and the winds and rains began to wear it down, the island began to subside, its weight cracking the earth's crust in a process known as isostatic sinking. Over time, Rangiroa Island sank beneath the living and dying bodies of trillions of coral reef animals, eventually sinking right through the crust. One day it was gone, and only the *pito* (Tahitian: bellybutton) remained — the coral atoll, marking the place where the inner earth once connected through the umbilicus of a magma tube to the outer earth.

Nor is the transformation complete, as the old Rangiroa Island sinks closer to the molten magma core from whence it came, preparing to recycle itself back into the shimmery, iridescent alkali metals that give the lavas of midoceanic hot spots their distinctive rainbow hues.

3

The Manysided Lagoon

K NOW THAT THE WORLD is uncreated," says the Jain *Mahapurana* (Great Legend), "as time itself is, without beginning and end." So too with volcanic islands, which cycle in and out of being through the dark compresses of the earth's interior. In Jainism, a philosophy that arose in India at the time of Buddhism, the world is seen as a place of unimaginable subtlety, where everything — not only volcanoes — is first here, and then invisible, and then there.

This manifestation of nature is embodied in a point of view the Jains call *Anekantavada,* or the Doctrine of Manysidedness, which suggests seven predicative states (Western thought, as far back as the Greeks, claims only two, as in Socrates is either a mortal or not a mortal). For Jains, all reality is true only from the perspective of the judge.

Staring out at Rangiroa's lagoon when the sun is setting, you might imagine a simplified version of this multifaceted view of reality. The setting sun is gilding the ripples on the surface of the lagoon. From your perspective, the ripples are evidence that the wind is blowing. From an underwater perspective, however, the gilded ripples are evidence that schools of mullet are browsing. A third possibility incorporates the two views: perhaps the ripples are the wind blowing *and* the mullet. Both are true. A fourth point of view from the Jain perspective is that because the ripples are the wind *and* are the mullet, then the true character of the ripples is *syadavaktavya,* or indescribable. And so on. *Ane-kantavada* itself is composed of two related doctrines: *nayavada* (viewpoints) and *syadvada* (maybe).

Rangiroa is a good place to entertain the concepts of viewpoint and many and how they combine to form manysidedness. To the ancient Tahitians, arriving in the Tuamotus somewhere around 950, the lagoon was the abode of the gods. To Magellan, in 1521, it was a new land that was disappointingly not the Spice Islands. To the Catholic missionaries arriving in 1851, the lagoon, with its two passes, was a safe anchorage for the ships that would carry the copra their converts would produce. To scuba divers today, it is the underwater equivalent of an African big-game safari.

Yet Rangiroa offers ever deeper and more subtle variations on reality. For instance, if not for its coral reefs, all traces of the ancient Rangiroa Island would be long gone and we might not have known of its existence. Such is the fate of volcanoes lying beyond the Darwin Point — north of the Tropic of Cancer or south of the Tropic of Capricorn — where the water becomes too cold for corals to grow fast enough to keep up with the island's subsidence. Without the buffer of living reefs, extinct volcanoes in temperate and polar seas face the full brunt of oceanic waves, which shear off their summits as they sink. The resulting flattops, known as guyots, lurk below the surface. (No one knew of the existence of guyots until World War II, when a U.S. Navy commander, playing around with the newfangled echo sounder aboard his transport ship, discovered a chain lying beneath the waves of the Pacific.)

But in the subtropical and tropical waters between the horse latitudes — the critical zone between thirty-five degrees north and south — where the water temperature hovers around eighty-six degrees Fahrenheit, corals join lava to form one of the most prolific construction teams on the planet, producing islands with phenomenal productivity and longevity. Metamorphosis from a high island to an atoll takes eons. Midway Atoll, at the far north-western end of the Hawaiian Islands, is the remains of a high island formed when a hot spot visited about twenty-seven million years ago. Likewise, the Tuamotus' ancestry dates back about two hundred thousand centuries.

If you know how to use it, the serene view of Rangiroa's lagoon at sunset can tell you much about this metamorphosis. Long ago, when the high island stopped producing hot lava runs, corals took off in force, seeded by the never-ending supply of transparent larvae traveling the ocean currents. Settling out of the plankton realm, these new residents made homes in the sharp and shiny nooks and crannies of the island's cooling flanks. The hardiest species — those able to withstand the battering of open-ocean waves — took hold first. The reefs they built provided shelter on their leeward sides for the more delicate corals to colonize.

With the rains falling and carrying away the extinct volcano(es) pebble by pebble, the outline of the high island began to recede. Yet the coral reef, firmly established, held to the original boundaries. And so the pattern was established: as the volcano eroded and sank, the barrier reef held firm, while the sea flooded into the missing landmass, creating the lagoon. Throughout the ensuing eons, the island got smaller and the lagoon got bigger and all the while the reef maintained essentially its original dimensions.

The passing years brought storms, and some years brought cyclones, whose waves tossed coral rubble and coral sediment (sand) onto the top of the reef above the waterline. Gradually these broken bits of *Acropora* and *Porites* and *Pocillopora* corals were cemented in place — thanks to wandering seabirds who stopped to rest their weary wings on the coral rubble and left their droppings behind when they took to the air again.

The birds' phosphate-rich guano interacted with rain to form a stabilizing layer of sandstone, and this sandstone in turn became the foundation for a concretelike layer known as beach cement. What began as a few small platforms of broken coral skel-etons atop the reef grew and merged until the first slender *motu* took shape. Eventually these islets underscored the outline of the sunken island in brilliant white sand — giving you a place to sit and mull Rangiroa's reflections on manysidedness.

By the time the last of Rangiroa's volcano(es) slipped beneath the waves, the *motu* had already been colonized with plants drifting in on the waves: the swaying, fruitful coconut palms, the bay ce-dar, the *nono,* the fragrant beach gardenia, and the airy forests of lettuce trees, which the Tahitians call *puatea.*

Sitting on the lagoon side of a *motu* at sunset, watching the rip-ples of water, you can observe the effects of these colonizations. Behind you, the tall canopy of the *puatea* is a favored nesting site for the dark tropical noddy terns the Tuamotuans call *kirikiri.* The male *kirikiri* carries the components of the nest to the fe-male, one *puatea* leaf at a time, then waits, head cocked, as she examines and either rejects the leaf, or glues it into place with a dab of her multipurpose droppings. Used this way, the guano-matrix of the nest is delicate and water-soluble, and so the tight, relatively rainproof canopy of leaves provides an ideal home for the noddies and their future children. Plus the view of the ripply lagoon outside their front door is, from the birds' point of view, something like a golden tablecloth. Or so I imagine.

In the Jain view of the world, everything has a soul. The *kirikiri* in the *puatea* trees have souls, as do the *puatea* themselves, as does the earth, including all its component stones, clays, min-erals, and jewels, as do fire-bodies, wind-bodies, and water-bod-ies, including the lagoons, oceans, rivers, and lakes. Surrounded by a soulfilled universe, Jains follow a path of intense asceticism. Even ordinary followers are prohibited from engaging in agricul-ture, animal husbandry, or forestry, and consequently, Jains domi-nate the worlds of banking, moneylending, and trading. Monks of some sects go naked, since the process of producing cloth harms many plant-, air-, and earth-bodies. They sweep the ground ahead as they walk so as to avoid stepping on insects, while wearing facemasks to prevent inhaling wind-bodies. Jain monks and nuns strain their drinking water and eschew bathing or swimming so as not to harm water-bodies.

Truly water is alive. And here at Rangiroa you get a sense of the fullness of that aliveness. Not just in the otherworldly colors and textures of the coral reef but in the *work* of water, the monumental tasks it undertakes and completes. Obviously, the formation of *motu* is one of them, the water tearing away living coral structures from their cement anchors and lifting them high above the reef to deposit in what will shortly be dry air again.

On the extreme northwesternmost *motu* of Rangiroa lies a jumble of huge boulders known as clasts. The biggest of these run thirty-five hundred cubic feet, the size of three-story houses, and are estimated to weigh 4.4 million pounds. They were deposited some three hundred years ago by tsunami waves, and today inhabit a *motu* otherwise devoid of topographical relief, looking as out of place as petrified dinosaurs. They invite you to walk first this way, then that, feeling the old, rounded sea urchin bore holes, the calcified barnacles, the remains of oyster shells. Then, having circumnavigated, you are invited to climb, using as handholds and footholds all the ancient topography formed by living things that were abruptly ripped from the sea in an event of unimaginable fury.

From the summit, you can see the powerful ocean marching in, wave after wave smashing itself against the shore. During normal tradewind weather, the surf is awesomely powerful, and its thunderous rise and curl onto the reef's edge is loud enough to force you to yell at your companions to be heard. But even five of these combers combined could not rip one of these clasts from the foundation of the reef, carry it into the air, and drop it where it now sits. For this, something monumental must have occurred in Alaska or Japan or some other latently violent place along the Pacific's volcano-and-earthquake-scarred Ring of Fire.

From the top of a clast, in the opposite direction, you can also see the lagoon, its water calmed by the breakwater of the living reef. With the sun behind you, the lagoon becomes a mirror reflecting blushing cumulus clouds. You are there too, your shadow spilled across the still water. And then a baby blacktip reef shark, who nurseries in this quiet corner of the atoll, ambles through, shedding multicolored ripples from its dorsal and caudal fins, as the scene breaks into the seven possibilities of predication . . . *Ripples that are wind?*

But what you are really likely to be thinking as you straddle the peak of the clast and gaze out to sea is that this coral dinosaur, quite likely the highest point in all the Tuamotus, would be the best place to run to if you knew another tsunami were coming this way.

4

Changeover

YANN APPEARS, AS SILENTLY as ever, and points at the base of a purple staghorn, where a school of raccoon butterflyfish, with their bandit-striped face-masks, flutter. A score of them are bouncing off the *Pocillopora* and off one another in a pisciform mosh pit, faster than my eyes can take in. Fish move at a superspeed (as do birds, most mammals, and every last insect) impossible for the unaided human brain to follow. If we had film of this fish dance and could slow it down and play it many times over, we would observe layers of behavior that in real time suggest only chaos. Yet Yann apparently recognizes more than I do, because his finger is pointing at something I can't see.

To discern what's happening in real time underwater it helps to look without attempting to see the actions of any one yellow fish in the middle of a jostle of yellow fish; in other words, it's easier to examine the overall pattern of motion. And what that is showing, with the help of Yann's pointing finger, is a small damselfish known as a black-spotted sergeant in the center of the mosh, wearing bold black-and-silver stripes. I know this is a male sergeantfish because he's guarding a purple cluster of eggs, and I can see that he is not enjoying the thrash but is desperately trying to shield his eggs from the butterflyfish whose offensive school is overwhelming him.

One by one the purple eggs are disappearing into the protactile mouths of the butterflies, delivering their salty caviar taste with a satisfying little *pop* (or so I imagine). The sergeant is frantic, darting and nipping in all directions at once. Although he manages to shred one butterfly's caudal fin, his war, sadly, is being lost. Other fish are streaming in: pyramid butterflyfish with their stunning tricolor geometry; a pair of halfmoon picassofish that look like cubist paintings; and a filefish sporting blue graffiti. Yann and I watch the attrition of the sergeant's nest from a one-foot-square mass to half of that, to half of that, to nothing. We could intervene. But who is to say that the butterflyfish-soul isn't entitled to snack on the sergeant-soul?

Yann's finger is still pointing and I know there is another layer to this struggle, though I can't make it out while the death throes of the nest are under way. Only when the sergeant abandons his future and heads down the reef slope to a large plate coral do I see what Yann sees. Here, on the underside of the underwater equivalent of a massage table, the sergeant is sloughing off his despair at the touch of a cleaner wrasse. Flaring open his gills, he is allowing the blue-and-black-striped wrasse (*Labroides bicolor*) — a four-inch-long fish with horizontal blue and black stripes — to poke its head into some of his most intimate places and nibble them clean.

At first glance these ministrations of a sharp-toothed carnivore might seem a penance — especially as the blood-red membranes of the sergeant's gills are actually streaming blood. Yet the cause of those wounds is what Yann has been seeing all along: an infestation of gnathiid isopod larvae — tiny crustaceans distantly related to lobsters, who rise from the sea floor at dawn, mosquitolike, to feed on the blood of fish, ticklike. They are the most common ectoparasites on reef fishes, and the preferred food of cleaner fishes.

Perhaps it was the swarming of the gnathiids that initially triggered the sergeant's distraction and prevented him from guarding his eggs with the usual vigor. Or perhaps their swarming made him flail in discomfort, thereby revealing his hidden nest site at the base of the *Pocillopora* coral. Perhaps it was both.

We have yet to see any hammerhead sharks, and at the end of this dive Yann goes from the regulator in his mouth to a cigarette in his mouth with more speed than usual. Somehow he can light a smoke in an open boat in strong winds while sitting beside the five-gallon gas tank for the outboard, dripping water from his sun-shredded hair and his threadbare wetsuit, as crew members hoist themselves aboard. It's a skill born of long experience and, today, something like frustration. He wants us to see and film hammerheads not because it would be good for his business (though it would), not because he would like to see them (though he would), but because he wants *you* to see them. And now, because time is running out and we are not seeing them, his frustrations are mounting. I tell him that it doesn't really matter, and he shrugs and rolls his eyes, palms upward, cigarette between nicotine-stained fingers — the classic Gallic *I don't care.*

Once, during the slow diving season at Rangiroa, Yann and his girlfriend visited me in northern California. They had been gone from the Tuamotus for less than a week, yet Yann — always fidgety and nervous in the topside world — had become as torpid as a hummingbird in the snow. I could hardly recognize him, as if I were watching a slow-motion playback. I thought the best thing was to get him out to the ocean and drove them to the chilly mouth of a nearby river, where harbor seals haul out to bask and white sharks frequently roam the big surf. Yann listened to my tour guiding, staring out against the spring winds that were making our eyes tear, jacket zipped to the neck, a hand-rolled smoke dangling from his lower lip. His girlfriend was thrilled to be off a *motu* only twelve miles long, but Yann was utterly passionless. No harbor seals, no white sharks, no cold water could inject his spirit back into that listless body. Only Rangiroa could. They would be traveling for another five weeks, and I actually wondered if Yann could survive it.

Back in Rangiroa, I try to explain that it doesn't matter if we find hammerheads, because the film will be what it will be. He shrugs again as if he doesn't care, but on the way back to dock he directs our boatman to stop at a couple of other likely places, where he pops overboard for a look. I tell him that my secret hope is to make a film about sharks without a single drop of blood in it — that it might, in its own way, contribute to the public education in manysidedness. But I'm not sure how well this idea gets across in my unschooled French.

One morning, on a crepuscular dive, when the dawn is just beginning to filter below, it comes to us. We are floating through subdued shades of blue and purple. In one of those fulcrum moments underwater, both the night shift and the day shift appear together, each defining the other, like shadows and sunlight. The result is a reef jam, as all the fish and cephalopods and crustaceans trade places, the diurnals relinquishing their nighttime hidey-holes to the nocturnals preparing to retire.

The onset of either day or night in the sea is akin to dropping a curtain and raising it again on a completely new drama. The familiar fish of the light or the dark retreat to beds among apartment complexes of branching corals, overhangs, caves, and crevices — beds only now being vacated by their counterparts, who are shaking life into sleepy fins and rousing themselves to go out and play their parts on the reef. The changeover occurs in the soft hour of twilight, when, for a brief moment, the daytime and nighttime species bump into one another — politely, or impolitely, according to their natures.

This morning, before breakfast, I am drifting sleepily along the dive site called L'Angle, watching the whitetip reef sharks who have been hunting all night sidewind across the bottom in pursuit of their sleeping caves. Turquoise clouds of chromis fish rise around them, snaring the plankton riding the incoming tide. Clouds of copper-colored sweepers mingle with the chromis, grabbing their last plankton mouthfuls before bed. The combination of turquoise and copper swirls hypnotically, and the reef, which I have seen many times in both the darkness and the light, seems twice as crowded, twice as colorful, as any time before.

Scattered across the slope, the cameramen are working their shot lists and keeping their eyes open for the unexpected. Although we are still hoping for hammerheads, no one diving Rangiroa this February has yet seen one. I am easily entranced by the little things going on in front of me, including an amusing exchange playing out in the small corridor nestled between an orange sponge and a blue *Porites* coral. Tucked into the corridor, an orangeband surgeonfish is still in bed. Meanwhile the diurnal occupant of the corridor, a sixspot grouper, is tired after a night of hunting and eager to climb into the same bed. I hide behind the bulk of the *Porites* coral and watch as the grouper swims back and forth in front of the bed, growing more annoyed until it darts forward at superspeed and inflicts what I imagine is a nip on the surgeonfish's head. Startled, the surgeonfish pops out backward and swims crookedly away.

I am watching the grouper settle into his tight bed between the orange sponge and the bluish coral, his colors perfectly matching his bedclothes, when some clue I am not conscious of makes me turn and look downslope. In the time that I have been watching this tiny element of the grand changeover, the sun has risen, and the world around me has been transformed into living color. A shoal of jacks shimmers below, parting into a pair of chrome-colored parentheses, framing the passage of a very large, barrel-bodied hammerhead shark. Swimming below and angling upward, it is stunningly enormous, so much bigger than the six-foot gray reef sharks that I find my brain trying to calculate and compare, imagining, through my adrenaline, that it is twice their size. I hope one or both of the cameramen have seen it, but I can't bring myself to tear my eyes away and verify that.

Hunkered behind the *Porites*, I watch the hammerhead flow over the cropped corals, its dorsal fin riding tall. The hammer itself is grotesque and fascinating. As the shark swims by at an oblique angle, I have the opportunity to see, out on the tip of one hammer, a dull black eye swiveling forward and backward, up and down. This translates in my mind as a strangely intimate glimpse into the shark's mind, its life of perpetual, detached searching.

And then it's gone, sliding over a rise and along the outer reef slope, backlit for a moment before disappearing. When I look up, I see one cameraman lined up on it and swimming hard. But already I'm imagining that a downward shot at this time of day is unlikely to yield more than a gray shadow against a gray background. Another aspect of *anekantavada*.

5

Breath Control

I T CAN TAKE YEARS of experience to begin to see even a small part of all the reef has to offer. And see is about all we can do, although a wealth of information is adrift in the ocean from electromagnetic fields, subsonic noise, subtle changes in water pressure, and chemical tastes and smells. The ocean's residents possess an array of specialized organs that enable them to read the signals hidden to me.

Yet, drifting over the reef, my senses feel full, sometimes overwhelmed, by the wild color and sudden motion, by the muffled watery sounds, by the joyful balance of near weightlessness. So much so that I know that even those things that I *can* sense frequently escape my notice simply because there is so much of it going on around me in three dimensions at all times. Then I try to imagine the worlds hidden to me: the smell of a fish's anxiety, the taste of water filtered through a sponge, the sound of a whale calling from hundreds of miles away, the feel of pressure waves emanating from a nudibranch climbing a fan coral.

If you're observant, you can use the reef's inhabitants to augment your own muted senses. The humphead wrasse (*Cheilinus undulatus*), who the Tahitians call *mara* and the French call napoléon, is an imposing fish up to seven feet long and four hundred twenty pounds, with an overhanging forehead, thick lips, and a blue body overlaid with squiggly patterns of green and yellow that look like the insides of a circuit board. This design is most striking around his eyes, and as with many reef fish, continues right into his eyeballs. He is a stately, easygoing fellow, whose leisured life dining on mollusks and poisonous things like crown-of-thorns starfish invariably allows him enough free time to greet you on your visits. Follow him and you'll find things you might not see, including every last cleaning station within his ten-thousand-square-foot (or more) home range.

You won't have to spend long on a napoléon tour of the reef before concluding that he is a hedonist of the first order, a sucker for all the ministrations of little tickly things. Over there, in the cool water under the parasol of a large plate coral, he rests for a few moments while an army of small, daintily colored shrimp crawl inside his lips and around his undulating caudal fins. They work like fiends, tiny claws picking away who-knows-what and frenetically conveying it to their mouths. The humphead hovers in a state akin to meditation. Only now and then he flinches and the shrimp somersault off into midwater. Someone pinched him too hard. Never mind. He won't hold it against them. He waits patiently until they climb back aboard and resume their body-work.

But was that enough? No. It's midmorning, that breakfast of supposedly inedible boxfish and sea hares is apparently weighing well in his belly. Time to visit the cleaner-wrasse shop at the corner of what might best be described as Drop-Off Way and Lobe Coral Alley. Although there is already a line of fish waiting for the tiny *Labroides bicolor* (a distant wrasse relative of the napoléon), our tour guide jovially jumps to the head of it. No one is big enough to contest him (though, in all fairness, I have often seen napoléons waiting in line, too). Opening his mouth as if to a dentist, he reveals his oversized canine teeth, and the cleaner eagerly enters and begins to floss. All the while the napoléon holds so still that he appears to fall into a trance of pleasure, losing his balance and drifting down and off to the right, while falling over on his side. The cleaner works around this, darting in and out of the pikelike teeth, ignoring his client's enraptured bumpings and bouncings off the pillars of lobe coral.

A wonderful series of papers by an evolutionary biologist in the UK named Redouan Bshary explains the complicated relationship between cleaners and their clients. In the first paper, Bshary, at the University of Cambridge, and his colleague Manuela Wurth from the Technical University of Munich, studied the bluestreak cleaner wrasse (*Labroides dimidiatus*) in the Egyptian Red Sea. Apparently, cleaners do occasionally bite their clients too hard (some actually sneak a forbidden fish scale now and again), which can lead to them being chased away by the angry client. If the injured client returns later to this same cleaning station, the cleaner will spend the first few moments backing gently into him and massaging him with undulating caudal fins. The client appears to enjoy this, and the researchers conclude that it is an attempt at reconciliation — an apology, if you will, and the first of its kind seen in anything other than mammals on the order of primates, dolphins, sheep, goats, and hyenas. Bshary also discovered that cleaners are more likely to use caresses to mollify carnivorous fish who might eat them.

But what of the nonpredatory client species, the vegetarians who can't wield the ultimate threat? What is to stop the cleaners from nibbling their fish scales, or nipping a little muscle tissue, or slurping an energy-rich slick of mucus? In another study with Daniel Schaffer of the Max Planck Institute for Behavioural Physiology in Seewiesen, Germany, Bshary discovered that market forces shape the cleaner-client relationship. Curbing what might otherwise be an overwhelming temptation to imbibe some of the client's protective mucus is the pressure of competition. Clients are less likely to visit cleaning stations where they have previously been cheated by the cleaner's nibbling, or where they have had to wait in line. So even nonpredatory clients wield the big stick of consumer choice.

Bshary's research also shows that cleaners treat roamers and residents differently. Residents with small territories and access to only one or two cleaners are more or less stuck with their local business owners. Roamers, however, travel far and wide and can pick and choose between the best cleaners, are more likely to be ushered to the head of the line, and are robbed less often of scales or mucus.

From the cleaner's point of view, the napoléon bouncing off the coral in his technicolored suit is a large and parasite-rich roamer, and therefore more important than the local vegetarians waiting in line. Yet all clients are worthy of notice, since virtually everyone visits a cleaning station occasionally, if not frequently. Lexa Grutter of the University of Queensland observed single cleaner fish servicing up to twenty-five hundred reef fish per day and consuming twelve hundred parasites in the process. Some clients visited as many as one hundred fifty times a day.

Most cleaners hang their shop signs on obvious promontories on coral heads, flashing their blue or yellow or black stripes by swimming in characteristic up-and-down motions in semblance of aquatic barbershop poles. Clients also find cleaners through means not available to human senses. In another study, Justin Marshall, working with Lexa Grutter, discovered that cleaners share an unusual color they call cleaner-blue. Characterized by a long wavelength not visible to the human eye, this blue is different from normal fish blues yet is commonly found in cleaners from a wide range of fish and shrimp families in the Pacific and the Caribbean.

As the cleaners are sending out signals advertising their place of business, the fish waiting in line are sending signs of their own, adopting strange, frozen postures to indicate their readiness to be cleaned (an important indicator if you are a large carnivorous fish who might otherwise eat little fish, lest the cleaners refuse you service). These poses include yawning, or waiting with the mouth open, as well as hanging upside-down or head up. It's one of the reef's more amusing harlequinades to come upon these fish of all sizes and stripes and colors miming dead fish or disabled fish or just plain boring, nonthreatening fish. Ever willing to help in their own care, some fish with bright or confusing colors blanch at the cleaning station to help the cleaners locate their pests.

Labroides, like human barbers of old, also practice surgery and dentistry, and the fish queue often includes clients with chunks of flesh missing from flanks, shredded fins, or putrid or infected wounds. You can even experience this debridement yourself if you approach quietly, holding out some battered part of yourself — usually your hands, which, after repeated diving and hauling yourself in and out of the Zodiac, are bloodied and infected. It doesn't take long to understand what it is that brings the clients back: a feeling like having your teeth cleaned, both pleasant and unpleasant, but one that you know is good for you.

Your chances of experiencing this are improved if you practice *pranayama,* the yogic breathing meditation. You might try slowing your breathing, making it very gentle, then holding your breath between each inhale and exhale (being careful to maintain buoyancy control). Because *pranayama* minimizes the exhaust from your scuba, it helps the little *Labroides* see through the bubbles of your alien being to the tissues in need of treatment.

The feeling of being tended to by a cleaner wrasse is different from the feeling of being bitten by a fish who mistakes your fingers for food. Whereas the latter are sharp attacks that take you unaware, the former are surprisingly gentle ministrations — though the action with a cleaner is often too fast to see. This lag, however, enables you to study the interesting delay between what you see and what you feel, because only after the *Labroides* has darted in and is backing out at superspeed do you notice the strange, tweezing sensation, as if the fish is plucking hairs where you have none. Now and again you may feel pain where it has bitten into living flesh. But the cleaner seems to understand a flinch in any language and quickly follows up with nibbles as tender as kisses.

Yann is drifting in his characteristic pose, arms wrapped around his chest, body canted to the side, as if reclining on an invisible hammock. His breath is as ephemeral as spider's silk, his silver bubbles marking the contours of the current. Slowly, casually, he drifts into the center of a group of surgeonfish, who are spawning again this night.

The spawning is a fast, jerky, staccato performance — the bullet-like upward flights, the release of milky spawn, the plummet back to safety — all swirling around the still center of a reclining, almost unbreathing Yann. It's tempting to try to join him, to experience, in an anthropological way, what the heart of the ritual feels like. But I know that whatever it is Yann possesses, or has trained into himself, or has stolen from the fish world, is not available to me, and I would rightfully be greeted as a gate-crasher.

Yann is not the only outsider in attendance. Mixed schools of anthias — diminutive, six-inch-long relatives of groupers, wearing primarily purples, oranges, and yellows — are darting hyperactively through the scene. Jumping, stopping, backing up, starting again, they are changing pace and direction as if running into glass walls. But there are no such obstacles in the ocean, and what these little planktivores are really doing is hunting the juicy, protein-rich eggs of the surgeonfish.

Alongside the anthias, a squadron of pretty, elongate speedsters known as bluestreak fusiliers has appeared and is trying to muscle the anthias out of the action. But the latter are too small, fast, and unpredictable to be intimidated. Meanwhile, the surgeons make no effort to defend their gametes against the encroachers, relying on sheer numbers — perhaps thirty thousand eggs per female per day — to overcome even the best efforts of the protrusible mouths, the endless appetites of the reef.

I am inching along the bottom, trying to get closer without disrupting anything, the ultimate voyeur. No one seems to notice. By staying low and breathing even lower, I am as good as irrelevant. In the falling light, the spawners are silhouetted against the twilight of the surface. They spiral around the reclining silhouette of Yann, ascending the ladder of his knees, elbows, and shoulders into the open water above. The anthias dart upward with them, bouncing off the imaginary glass ceilings as they go. The fusiliers zigzag in a school as tight as a sheath of arrows.

Then a newcomer appears, a couple of newcomers, in fact. From their characteristic bobbing, I realize that a pair of cleaner wrasses has joined the fray, performing their barbershop pole routine, trying to sell their services. Amazingly, there are takers, including a few surgeonfish, who abandon the orgy for a tickle and a nibble. It's only a quickie, apparently — and nearly as fast as they touch down, the surgeons streak back to the commotion above, ascending Yann's angles to reach the exploding fireworks of spawn and eggs as the current rips by above.

Cleaning stations are widespread and common on coral reefs, with more than fifty species providing services, including six shrimps, one crab, and one worm. Clearly the need for cleaning and surgery is great, and some long-distance travelers, such as the pelagic turtles and tunas, visit the coral reef specifically for cleaning and/or medical services.

Cleaner fishes usually cover the day shift, while shrimps specialize in cleaning shy customers in dark crevices or at night. These shrimp are primarily of the genus *Lysmata* or *Stenopus*, dainty of limb and gaudily colored, and for the most part they work in lifelong mated pairs. This tendency for fidelity extends even to their clients, with some shrimp pairs joining moray eels, large groupers, or napoléons in their home caves and cleaning them exclusively or nearly exclusively. Such valet shrimp often faithfully follow their patrons when they move to a new home. For oversized clients, a resident pair of shrimp offers an important health benefit, since some shrimp not only clean the exterior of the skin but also make small incisions with their nippers to excise subdermal parasites.

Happy to clamber among *your* teeth and tongue, the shrimp cleaners will check out the crevices of your nostrils and around your eyes, if you're willing to remove your mask and breathe as lightly as a flutter, while the multiple tickly legs and pincers explore you. As Rumi advised: Don't let your throat tighten with fear / Take sips of breath / all day and night.

If you spend long enough in the company of cleaners on the reef you might even find yourself a victim of one of the inevitable opportunists: the false cleaners, or mimic blennies, small fishes from the genus *Aspidontus* or *Plagiotremus,* who make their living as cleaner mimics (some only in their juvenile phase). These little predators look uncannily like cleaner fishes, with blue stripes and long, narrow bodies. They even abandon their wriggly blenny swimming style for the cleaners' characteristic rowing with the pectoral fins.

Even more remarkable, some false cleaners have stolen the barbershop-pole display, the trademark of the cleaner wrasses, to lure the unsuspecting and the disheveled close by. When a victim falls for the ruse, the false cleaner charges, bites off a hunk of flesh with fangs hidden in its underslung lower jaw, then runs to hide in a cranny, usually an empty worm tube in the coral. The scale-eating blenny is renowned for darting up from the substrate and nipping at passing divers.

Studies have shown that juvenile fish are most prone to the deceptions of false cleaners, and that only painful experience teaches them the difference between the real thing and the false. Justin Marshall at the Sensory Ecology Laboratory in Queensland has found that false cleaners do not wear cleaner blue (nor do juvenile real cleaners, which is confusing) — and this is apparently one of the lessons young fish must learn on their way to a hygienic adulthood.

6

Inside the Turtle's Shell

T O FULLY EXPERIENCE the underwater world it helps to practice *pranayama*, which resembles the technique used by all air-breathing mammals in the sea — and nearly exactly mimics the breathing seals use when they come ashore. At the mouth of a river in northern California, not far from where I live, a rookery of harbor seals (*Phoca vitulina*) hauls out to groom and to sleep. Nostrils opening wide, each seal inhales a sharp, audible *hisssh,* which shuts off abruptly as the nostrils snap closed. The seals' breath-hold, which biologists refer to as spontaneous apnea — or if the seals are asleep, sleep apnea — lasts between one and four minutes. Even on dry land, the seals continue to breathe as if in the water, and though four minutes may seem a heroic length of time to hold your breath, it probably amounts to rest and relaxation for a harbor seal who spends his or her life ascending from depths of three hundred feet after seven or more minutes underwater.

By practicing a human equivalent, you will find that animals underwater are drawn to you during your inhale and breath-hold phases, then drift farther away as you exhale, then return as you breath-hold — creating a pendulum of swaying fish moving to and from your center. This imparts the illusion that the fish are actually attached to your breath, as if they live on the balloon of your breath, and are moving away and moving closer as you breathe. In this way, even without an active *pranayama* practice, scuba-diving provides a platform to become more intimate with your breath and its effects on you and the world around you.

Above a dive site called Récif d'Avatoru, outside Rangiroa's other big *hoa*, the Zodiac idles, as we fall in slow motion toward the spur-and-groove zone. This is the part of the reef where oceanic swells crash ashore carrying sediments, some of which are deposited on the *motu*, some of which tumble downhill underwater. The scouring action of the tumbling sediments combines with the powerful backwash of the swells to carve grooves in the reef. While the tougher corals and the coralline algae make a living on the spurs, the grooves are stripped as bare as avalanche chutes.

Yann is carrying the morning's props, a couple of tuna heads and the better part of a tuna back. We follow him down a sediment chute, behind a gigantic and growing school of locals who follow his tuna scent: the resident butterflyfish, filefish, goatfish, Moorish idols, triggerfish, porcupinefish, and wrasses of every color and description. Yann is a veritable pied piper, and before long small blacktip reef sharks join the parade, happily jostling fins with all the little fish. Even stingrays rise from under the blanket of the sandy bottom and flap alongside. When Yann disappears over the lip of the outer reef slope, the flowing, multicolored scarf of fish, now thirty feet long, tumbles after him.

We set up shop on the first terrace on the wall, a sandy platform about ten feet wide and eighty feet down, and stake the tuna scraps to the outer edge of the terrace. The idea is to spike the stream with the irresistible scent of tuna oil, in hopes of attracting the heavyweights of this pass, the silvertip sharks (*Carcharhinus albimarginatus*). But until then, we need to fend off the hordes of nibblers trying to dismantle the chum molecule by molecule. Gently, we kick at the fish with flexible fins. But this only stirs the tuna scent. Far better to breathe on them, exhaling voluminously from our regulators. I lean out, breathing hard, which momentarily disrupts the attacks on the chum, although a few opportunistic thieves sneak behind and nip at my hands, taking skin and drawing blood.

The arrival of a pair of silvertip sharks doesn't disturb the nibblers at all, even though the sharks swaggering up the wall are eight feet long and heavyset with muscle. They wear the navy-gray of their fellows in the family Carcharinidae, which in their case is highlighted with white tips on the pectoral fins, the caudal fins, and the trailing edge of their dorsal fins. At this depth, and in this light, their colors appear impressively dull, almost anonymous. But when we switch on our filming lights, tethered by a couple of hundred feet of cable to the generator in the Zodiac, the sharks' luster ignites, their grays shining like polished nickel, their whites gleaming like platinum.

Despite their tough appearance, the sharks approach warily, circling from afar. Contrary to public opinion, most sharks are cautious animals, particularly while scavenging, when they are attentive to the fact that they will likely meet members of their own dangerous kind on the other end of the same scent trail. To entice them, we lighten our breath, inhaling and exhaling slowly, staying still. The clouds of little fish take advantage of our hiatus and swarm the chum, all their colors and shapes orbiting like electrons around a nucleus. The sharks tack back and forth, gaining speed as they gain confidence — until, with a shiver of her tail, the larger female darts forward, bulldozing through the hordes, and emerges with a tuna head in her underslung jaw. Many fish pursue as, catlike, she shakes the chum between her teeth, everts her lower jaw, and engulfs the meal whole.

Without realizing I've been holding my breath, I release it, and the bubbles scatter the remaining fish off the second tuna head, opening a path for the male shark. He comes in low to the ground and, without breaking stride, dips his head, scooping the chum off the terrace. There is enough slop for everyone to get at least a taste, as the shark beelines downslope, dropping bits of tuna and towing a net of fish behind him, including one of the stingrays flapping hard in the open water where they rarely go. I hold my breath as the parade departs, listening for the sounds of their motion, something like the whistling wings of a passing flock of geese, I imagine. But even holding my breath, I cannot discern such delicate echoes in the ocean.

In the dry world, the practice of *pranayama* is designed to clear all obstacles that prevent the body from being fully inhabited with the vitality that makes life — what the yogis call *prana*. One of the first and continuing lessons in a *pranayama* practice is to become conscious of the breath, to rescue it from its autonomic hinterland. On scuba, however, your breath ceases to be an involuntary function and takes center stage as *the* thing keeping you alive, being the loudest thing in your universe, the ultimate Darth Vader breath. It does this through the regulator, the mechanical device feeding compressed air from your scuba tank into your mouth.

Individual regulators have a distinctive feel to them — hard-pulling or soft-pulling, for instance — but all deliver an effect different from breathing on land. When you begin an inhale through a regulator it feels as if you are encountering resistance, a hard knot that the breath can't get past. This quickly releases or dissolves as the sensation of pulling hard is suddenly replaced by a forced inrushing of air. Depending on the quality of your regulator, these sensations can be strong or subtle but they are always present. The inrush seems to inflate your lungs faster than a normal breath, and, whether you practice *pranayama* or not, there's a tendency to hold this air in your inflated lungs for a little while before exhaling, as if it came in too fast to get all the oxygen out of it and you need to sit with it a while.

Of course, since your first day of scuba lessons, you are warned at great length about the dangers of holding your breath underwater, because air compresses further as it is subjected to the pressures below. The behavior of air inside either your scuba tank or your lungs is an example of the physics of Boyle's Law, which states that at a constant temperature and mass, the volume of a gas is inversely proportional to the pressure exerted on that gas. This is not a problem during your descent, because as the pressure is doubled, the gas you are breathing is halved in volume. But it becomes dangerous on the ascent if there is no outlet valve (your exhale) for the reexpanding gases in the closed container of your lungs. If for some reason or another you hold your breath and ascend even a little, you may suffer a pulmonary barotrauma, more commonly known as a burst lung, characterized by hoarseness, shortness of breath, and chest pain. Or worse, you will suffer an arterial gas embolism (AGE), which occurs when bubbles from the burst alveoli (the air cavities in the lungs) leak into the bloodstream and travel to the brain, mimicking a stroke. The onset of an AGE is sudden, the prognosis is frequently poor, and the only real hope is rapid treatment in a recompression chamber, which, unfortunately, is usually far from the coral reef.

So you are coached to breathe in and breathe out and to never hold your breath. Even in an emergency ascent — a maneuver designed to get you to the surface when you've run out of air — you are instructed to hum on your way up, even if your lungs feel empty, because there will be lingering pockets of air that will expand on the ascent. Humming and ascending no faster than your bubbles, your urge to breathe will dissipate as those last bits of air in your lungs reexpand. By the time you reach the surface, you may not feel much of a sense of emergency about breathing anymore.

And this is another thing that entering the underwater world does for better or worse: opening the gates of manysidedness to interpretations of emergency. In free-diving (breath-hold diving without scuba) you also experience the feelings of empty lungs down deep, which then magically reinflate, if only a little, as you ascend. At the lowest point of your free dive, you might experience your lungs beginning to uncontrollably clutch for air, only to have that feeling disappear on the way up. After gaining some proficiency, you learn to weather the panicky feeling accompanying the clutching.

Yet the line between safety and recklessness is as permeable as the surface, and about the time you have learned to withstand the clutching, you may have also discovered that a little hyperventilation through your snorkel at the surface increases your dive time. Perhaps you discovered this innocently enough: returning to the surface after a dive that pressed your limits, breathing hard there, and then seeing something of unusual interest below — a juvenile batfish mimicking a flatworm, perhaps — and diving immediately in search of it. Miraculously, you stayed underwater longer than usual with less urgency. From there, it may have required only a short course in trial and error until you learned that hyperventilating before a dive, and diminishing the amount of carbon dioxide in your lungs, rewired the breath-hold breakpoint — that threshold at which you say, *I must breathe.*

It's a wonderfully exhilarating realization that a little heavy breathing at the surface will transform you into a master diver. But your sense of emergency is once again skewed by the underwater world since you are lying to your brain, convincing it that your need for air is not so urgent. In fact it is as urgent as ever, and in bypassing the signals you risk suffering an underwater hypoxic loss of consciousness, or HLOC — better known among divers as shallow-water blackout (unconsciousness usually occurs close to the surface). This is the leading cause of death in free divers, and has claimed some of the sport's champions. Its effects are so insidious that even advanced practitioners have died while training in the shallow end of swimming pools.

Although you are admonished never to hold your breath underwater because of the risk of arterial gas embolism, you can do so, and will do so, though preferably in a conscious way that does not compromise your longevity. The reward of carefully holding your breath underwater is to hear the sound of the world without your breath in it, which is not, despite your fears, the sound of death.

By leaking the breath, you may be rewarded with an observation of a school of bumphead parrotfish (*Bolbometopon muricatum*), normally among the most wary of fishes due to the human predilection for eating them. Up to four feet long, and one hundred seventy-five pounds, bearing large humps on their foreheads (Australians call them the double-headed parrotfish), these monsters roam the reef in bisonlike herds up to fifty strong.

Bumpheads feed primarily on living coral, and their bite marks — large white semicircular gouges, often aligned side by side — impart a characteristically strip-mined look to a coral head. On occasion they ram the coral with their humps, backing up and advancing headfirst in short but violent sprints. The effect is stunningly destructive, as chunks of coral fall away. The noise is memorable, too, like a wrecking ball wrapped in neoprene hitting hollow brick. The combined soundtrack of a school of ten bumpheads ramming coral is akin to a percussion orchestra drunkenly tearing down the house.

If your practice of underwater *pranayama* is proceeding well, you may find that the balloon of your breath is growing stickier — bringing the creatures of the sea closer to you. This gives you the opportunity to learn that what you hear underwater, and the way that you hear it, is not the same as in the topside world. In the air, sounds reach each of our ears at slightly different times and in slightly different amplitudes, enabling us to discern direction. But in the water, sounds reverberate across the skull everywhere at once, destroying our directional acuity and nearly erasing our ability to hear higher-frequency sounds.

Reduced to a low-frequency soundscape, the underwater world sounds like the slow motion playback of a tape recorder with failing batteries. Seals have dealt with this by evolving inner ears buffered from most other bones in the skull, allowing them full directionality and auditory fidelity.

With our ears full of water, with no idea from which direction sounds are coming, and barely aware of the higher-frequency noises, we might be tempted to ignore hearing altogether, as we concentrate on the seemingly more rewarding sense of sight. But then we would miss the treasures that rescue the underwater world from its one-dimensional fate.

Another option is to train ourselves to hear better, and to hear more, underwater, and modifying the breath will help. But beyond that we might explore the fifth limb of yoga, *pratyahara*, which is usually described as withdrawal from the senses, but which also means feeding the inner senses. *Pratyahara* is associated with the turtle withdrawing its four limbs, its tail, and its head (signifying the five senses and the mind) into its shell. In diving, we get the chance to experience some of the benefits of *pratyahara* without the learning curve, since we have already crossed at least partially into that quieter (not to mention voiceless) realm of the stilled senses.

As a novice diver, you might find this oddly stimulating, as your mind fights to make the world as clamorous as normal, inviting all the sights, colors, and peculiar motions into the dizzying whirl. But over time, breathing lightly, you have the opportunity to allow the natural sensory deprivation of the underwater world to expand your observational powers — to become the inside of the turtle's shell.

When that happens, the soundscape actually becomes more accessible. In fact we *can* hear the higher frequencies underwater, they just sound different than they do above, both quieter and more internal. Once you know how to listen, much of the soundtrack of the sea feels like it's happening inside your body. If you've ever heard the drumming of ruffed grouse in spring, a sound you feel in the chest rather than hear in the ears, you know what this is like.

7

The Near-Field/
Far-Field Boundary

ON A DIVE AT SLACK TIDE in Tiputa Pass, during that rare moment when the currents rest and the water lies so still you can see the sediments suspended in it like silver dust motes, the resident napoléon is doggedly following Yann and his stream of leaked bubbles. We are on the outer edge of the pass, teetering on the brink of the deep, which falls away into the purply-gray colors beyond the sunlight. Much of the bigger life of the sea rides this edge — the aquatic equivalent of a big-game trail — winding in and out of the deep coral canyons.

But from what I can tell, the napoléon is traveling along this edge simply because Yann is here — for reasons that may be as complex as the doctrine of manysidedness itself. Perhaps Yann's pockets are trailing the scent of the tuna head he stuffed in there yesterday. Perhaps the fish recognizes Yann from their countless dives together and enjoys spending time with him. Perhaps both. Or maybe the napoléon is curious about Yann's current occupation: banging together two pieces of skeletonized coral rubble.

Tilting back and forth on his tall axis, big eye swirling through its circuit-board design, the napoléon edges closer. The sound of the clanging coral is muted, as if happening behind many closed doors inside my own body. It's not really even noticeable at first, and then it is. If I pay attention, it seems to originate from somewhere near my temples, like an erratic pulse. Yet many little fish are being sucked in by the sound, arriving from all directions as if reeled in on invisible fishing lines.

Among these is a school of blue-green chromis (*Chromis viridis*) — stunning little damselfish wearing an ethereal aquamarine. Normally, the chromis hover a yard or two above the coral, fishing in the open for transparent zooplankton. But just now they are sliding down the water column toward Yann's hands and the sound of the rapping coral stones.

Rowing up from below, a spotted soapfish (*Pogonoperca punctata*) creeps into view. Squat and fat with white polka dots on a brown background, commonly mistaken for a grouper, the soapfish usually spends the day wedged headfirst into a crevice with only its tail hanging out — a strategy it survives thanks to grammistin, a bitter toxin coating its skin, which is both a hemolytic (dissolver of red blood cells) and an ichthyotoxin (fish poison). Nocturnal, secretive loners, soapfish rarely roam the reef in daylight. But here is one drawn from sleep and natural shyness to investigate the potential of this sound.

Meanwhile Yann keeps tapping, and the chromis have apparently slaked their curiosity and are drifting back to their feeding station above a yellow coral head. The soapfish lies on the bottom, the strange little goatee of flesh hanging from its lower lip quivering. When the napoléon comes too close, the soapfish sprints across the bottom and hides behind a ledge.

Yann taps, interspersed with moments of silence, and the noise is apparently an irresistible mystery to many of the creatures of the reef because now it has attracted a lionfish (*Pterois volitans*), a venomous relative of the stonefish. But whereas the stonefish looked like nothing more than an algae-encrusted piece of coral rubble, the lionfish is of indescribable beauty — a lacy, plume-bearing extravagance of a fish, complete with red and white and gold stripes along its body, and wispy fins modified to look like feathers (hence its other common name, the turkeyfish).

Like the stonefish, the lionfish is also a deadly presence on the reef, bearing glands at the base of its feathers, injectible through its quills, which contain toxins. The denizens of the reef know all about this poison and give the lionfish a wide berth. Its prey are herded into dead-end corners and narrow alleys by a fluttering dance of its wings — victims of a hypnotic entrancement at the sight of the feathers fluttering in slow motion.

So confident is this enchanter of its defenses that it engages in aggressive displays with creatures many times its own size. Drawn in by Yann's tapping, this lionfish advances now in full-feather display, intrigued but unsure, and deploying all its weapons just in case.

I am also listening to the sound of the tapping buried within my own body, while imagining how it sounds for the lionfish, who is hearing something entirely different. Because the density of fish bodies is nearly the same as the density of the water, they are "transparent" to the displacement waves of the sound. However, their fluid-filled inner ears, called labyrinths, which harbor floating otoliths (masses of calcium carbonate), are subject to inertia, lagging behind the movements of their bodies in the sound field and providing directionality.

The teleost (bony) fishes possess yet another organ of sound unavailable to me, a swim bladder (or gas bladder or air bladder). While not specifically an organ for hearing, the swim bladder becomes an important accessory to hearing and sound production in much the same way that our lungs aid us in breathing and speaking. A gas-filled sac in the front of the fish's body, the swim bladder primarily controls buoyancy through a process of inflation and deflation.

On the descent, as the increased water pressure makes the fish negatively buoyant, air is siphoned from the blood into the swim bladder. On the ascent, when the air in the swim bladder is expanding and making the fish positively buoyant, excess gas is diffused out the swim bladder, into the blood, through the gills, and into the surrounding water. Inflation and deflation happen subtly, continuously, and autonomically, enabling fish to maintain the energy-efficient state known as neutral buoyancy.

Human divers clumsily mimic this elegant process with the BCD (buoyancy control device), a kind of inflatable lifejacket hooked to the scuba tank by a low-pressure hose, through which we manually add or subtract air to make ourselves lighter or heavier. As any novice diver knows, buoyancy control is a challenge underwater, and most learners spend an inordinate amount of their time hovering far above what they would like to see or bouncing unpleasantly off it. Experienced divers learn to regulate the amount of air in their lungs to roughly mimic the effects of a swim bladder.

The lionfish, hovering in neutral buoyancy a foot off the bottom, pinpoints the sound of Yann's tapping with the compass of its otoliths. The noise is amplified by the bubble of its swim bladder, which pulses sympathetically with the pressure changes in the water, stimulating the fish's inner ear. Listening hard, I discern only a muffled pinging — though I can imagine, as the lionfish draws near, how the sound might actually be as loud and clear as wind chimes on the underwater breezes.

Years ago the staff of the Hôtel Kia Ora — the only hotel on Rangiroa Atoll in those days — turned the generator off at nine each night, leaving their handful of guests in the dark and in the quiet. Sleeping in *fare* — traditional Tahitian bungalows made of woven pandanus walls and palm-frond roofs — the sounds of the night were transparently evident. Short of sleeping on the beach, this was the most pleasant way to experience the voice of the coral atoll: the soft wavelets of the lagoon, the deep thump of waves to seaward, the rustle of palm trees, the raspy laughter of fairy terns grumbling their insomnia. Without air conditioning or motor vehicles, the conversation of geckos living in the *fare*'s walls became nighttime intimates, their soprano cluckings as soothing as hens at roost.

Inside this cocoon and its absence of man-made noise, the soundtrack of the lagoon became a constant — and not just the wavelets, but the nocturnal sounds of the hunting and chasing and dying of things in the sea, which translated through the surface as sudden slashings of water, complete with slaps and mini-geysers, and then the rippling back to calm and ominous silence.

Hot and sticky, beset by tiny saltwater mosquitoes, or with eyes tearing from the fumes of insect repellent, I spent a fair amount of most nights awake. Far better to rise and wander the lagoon edge, where the sea's bioluminescence provided an intermittent lantern, blinking on and off with the violence below, stirring ephemeral colors of blues and greens — the wavelengths able to travel farthest in water.

Accompanying these sights were many sounds — not only the splashing and slaps on the surface, but also the concavity of thumps, the grunts and strange watery scratchings, the *crick-crick-crick* calls of the noddy terns hunting by night. I let these encoded noises wash through my mind — first approaching, then here, then gone. The night sounds of the lagoon reconfirmed my understanding of the coral reef as a wildly inscrutable place. Here, on the moist edge, I learned to listen to the subliminal messages of its undervoice.

Practicing *pratyahara*, listening in the realm of the inner ear, or what I came to think of as the third ear, I discerned the collective voice of the coral reef, which came across as a low hum or a vibration, like a cat whose purring you feel only with your hand. At the time it seemed to be the combined voices of everything in the sea, on the reef, in the mangroves fringing the reef, in the seagrass beds, and on the sliver of the atoll, as if I had recorded all the individual sounds from these places on a two-track tape, layering them one on top of another, so that echoes of the underlayers remained even as nothing was recognizable. I thought at the time that this hum might resemble the sound of distant life in the universe — or what we might someday hear of it after its travels across unfathomable space.

Years later, I heard about a team of Japanese seismologists who had recorded and described a deep, low-frequency rumble present in the ground even in the absence of earthquakes. Existing at between two and seven millihertz, in an auditory realm far below human hearing, this noise was dubbed the earth's hum. Previous researchers had supposed it to be noise in their seismological data, and had laboriously filtered it out of their samples. The Japanese team surmised that the hum might be the sound of variations in atmospheric pressure drumming on the ground.

But Junkee Rhie and Barbara Romanowicz from the University of California, Berkeley, suspected otherwise, and began to filter the data in order to exclude everything but the hum. They searched roughly sixty earthquake-free days each year, tracking the direction of the hum and triangulating from two receivers to pinpoint its source. As they suspected, the origin was not on the land, but in the seas. The hum migrated seasonally, following winter from the northern to the southern hemisphere and back again.

They hypothesized that this was the sound of storm energy transferring via infragravity waves to the sea floor, where it was converted to seismic waves — in a sense, the conversation between the sea and the sky of a turbulent planet. Could this be what I heard on the edge of Rangiroa's lagoon on hot, mosquito-infested nights? The scientists would doubtless say no. The yogis would probably say why not? And the Jains would sum up the likelihood as maybe.

Because of the differences in density and compressibility between air and water, sound travels 4.8 times more quickly in water than in air, with a wavelength 4.8 times longer in water than in air. As a result, it takes more energy to generate sound in water than in air, but this sound propagates farther and faster.

Sound traveling through the water produces a pressure wave that provides information about the intensity of the sound. Information about the direction of sound comes from particle motion within this wave and is imparted in three ways: displacement (the distance each water particle moves); velocity (the speed of the moving water particles); and acceleration (the rate of change in that velocity).

Nearby sounds underwater occur within what is called the acoustical near-field and conform to what physicists call spherical waves. As sound travels farther from its source underwater, it eventually crosses the near-field/far-field boundary to a place where the ratio of pressure and particle motion become a constant. At this point the spherical waves become plane waves, with different acoustic properties.

The acoustical near-field is loaded with information about both sound intensity and direction, and this is where most sound happens underwater: the sound of fish tails beating in the water, the calls of fish, the mining of snapping shrimp, the burrowing of sea urchins. In contrast, few sounds have enough energy to produce compressional waves capable of reaching the far-field underwater — although those that do are memorable even to the untrained human ear: the songs of whales, the cries of seals, the voices of some dolphins, and, increasingly, the cacophony of our own world. This includes the sounds of motorboats, Jet Skis, the airguns used to fire high-energy acoustic bursts for sea floor mapping, and midfrequency sonar blasts used by the military to detect submarines and by the oil and gas industry to search for fossil fuels underwater.

Scientists are discovering that our growing soundprint in the seas is a source of new troubles for ocean residents. The sounds we make tend to be louder and more persistent than those made by nonhumans. The low frequency active sonar (LFA sonar) used by the military to detect submarines is the loudest sound ever put into the seas. Yet the U.S. Navy is planning to deploy LFA sonar across 80 percent of the world ocean. At an amplitude of two hundred forty decibels, it is loud enough to kill whales and dolphins and already causing mass strandings and deaths in areas where U.S. and/or NATO forces are conducting exercises.

Cetaceans die when LFA sonar is deployed, at least in part, because they shoot to the surface in distress, causing nitrogen bubbles to form in their tissues — the same malady known to divers as decompression sickness or the bends. Scientists from the Woods Hole Oceanographic Institution recently proved that sperm whales are capable of suffering the bends after analyzing their bones and finding them pitted with old lesions, indicating mild but chronic exposure to decompression sickness in the course of their deep-diving lives. Necropsies of beaked whales beached in the Canary Islands following LFA sonar exercises revealed the same acute form of the bends seen in human divers who ascend too fast to eliminate the excess nitrogen in their blood.

A few people have had the misfortune of being in the water near an LFA sonar test. A scuba diver off California reported disorienting lung vibrations even though the transmitting ship was more than one hundred miles away. Navy divers experimentally exposed to LFA sonar required hospitalization and treatment for seizures. A Hawaiian resident swimming with dolphins during an LFA sonar test became disoriented and nauseated enough that a physician diagnosed her with symptoms of acute trauma. According to her court declaration, the dolphins responded in a way she had never seen, hugging the shore and the surface, as if trying to escape the water, while vocalizing in distress. The Navy admitted that she and the dolphins had been exposed to one hundred twenty decibels of LFA sonar, only half its operational strength.

8

Eavesdropping

F ISH HAVE A THIRD method for hearing that has no equivalent in the terrestrial world. This appears as a system of canals and grooves known as the lateral line. Punctuated with nerve cells, the lateral line contains hairlike cells designed to detect water movement. This hearing system is sensitive to different stimuli than either the labyrinth or the swim bladder.

The lateral line enables fish to discern the low-frequency sounds made by moving objects in close proximity. Sensitive over the very low range between ten and two hundred hertz, it is ideal for locating swimming prey, whether zooplankton or large fishes. But even more remarkable, the lateral line system enables fish to hear the vortex trail left behind *after* a fish or lobster has swum by. In this way, the lateral line acts as a brief-lived time machine, allowing fish to hear where something was a moment ago, and discern which direction it was traveling at that time — as if everything that moved in the topside world left a contrail.

Of all the impenetrable distinctions between the underwater world and my world, few seem more profound than this. Imagine knowing where something was not because you hear it while it is happening and then turn to see it, but because you can see it after it has happened. Perhaps because of this, the sensory abilities of the lateral line system are sometimes referred to as distant touch.

Alone in Rangiroa's lagoon, floating on snorkel, I am rewarded by the approach of a school of reef needlefish (*Strongylura incisa*). Each fish is more than three feet long, tubular like a flute, with beaklike jaws comprising a quarter of its length. These needlefish are colored so silver as to nearly disappear against the mirrored underside of the surface — and this is their main hunting strategy, to camouflage themselves against that naturally confusing cusp between the air and the water, where both light waves and sound waves distort.

As we float, the wind blows, and ripples brush against the surface. Or perhaps it's a school of mullet browsing below. Whatever the cause, the needlefish lurk inside this refraction zone, lapping partially into view beneath a miniature crest, then disappearing into the miniature trough, their forms shattered into a dozen flittery shards by the blinking of the water. I study them as best I can, piecing together the fragments until some semblance of a whole emerges: a school of six needlefish, pointed into the ripples, big eyes casting up as well as down, using the reflective underside of the surface to increase their field of view.

When I flick a fin, they maintain perfect distance, never allowing the gap between us to close. So I wait, sculling with a fin or a hand, but only when they move away. In this manner I am directed around this corner of the lagoon in the style of a needlefish, going where they go and at the speed they go. We drift above the coral heads dotting the sandy bottom, each complete with its residents: a pair of butterflyfish here, a pufferfish there, a school of damselfish tending their algal garden.

No matter how I approach them, the needlefish know where I am, and even when I steal up along a blind sightline, they respond to my presence, keeping a distance they deem comfortable, around six feet. Gliding forward, bathed in a symmetrical flow field that becomes distorted when it nears me, the needlefish receive information about every nonmoving thing in their immediate vicinity. This batlike talent allows them to navigate the convoluted topography of the reef, to maintain position in a school, and to get around at night. For blind fishes who live in caves or at great depths, this hydrodynamic imaging is their primary sensory tool.

The differences between me and these fish could hardly be greater. Even as I struggle to collect the smallest measure of information, the needlefish are seamlessly wracking up multiple, multilayered impressions of me — collecting intelligence not only with their eyes, but also with their labyrinths, their swim bladders, and their lateral line systems. Flooded with data, they can see where I'm going and where I've been. In contrast, I grapple to piece together a fuzzy snapshot from the torn anatomy of surface refraction.

Humans live by light, which travels fantastically well in air and in the trillions of miles of the blackest vacuum of space, yet barely penetrates three hundred feet into the water. As a result, we know more about the surface of the moon than about the deep oceans; and precisely because the seas are largely dark, we mistake them for mysterious, when in fact they are as full of information as they are of water — much of which we cannot register let alone understand. We may not be well equipped to listen in, but the underwater world is an ideal conduit for sound, with the oceans divided into layers that speed or hamper sound's travels depending on temperature, salinity, chemistry, and pressure.

The topmost sun-warmed layer of the sea (sixteen to thirty feet deep) relays sound faster than the cooler layer below, while the area buffering these two layers acts as an acoustical shadow zone, where sound waves undergo refraction, just as light rays do at the surface, bending away from the region where sound travels faster (the warmer water) and toward the region where it slows (the cooler water). Discovered just prior to World War II, the shadow zone became an important part of naval warfare, providing a place for submarines to hide from the pryings of enemy sonar.

While temperature generally governs the speed of sound in the upper reaches of the ocean, at depth, pressure takes over. Below the water layer known as the thermocline, where temperature drops rapidly with depth, the speed of sound slows. At a point roughly half a mile below the surface — the exact depth varies with season and ocean — the temperature becomes isothermal (constant), and the increasing force of water pressure gradually allows sound to accelerate again.

In the years before and after World War II, the U.S. Navy identified a region of water straddling the bottom of the thermocline and the top of the deep isothermic layer that they called the sound fixing and ranging (SOFAR) channel. Popularly known as the deep sound channel, it proved a place where — in keeping with the laws of refraction — sound waves ricochet back and forth between the slower thermocline and the faster deep isothermic layer. The end result is a narrow yet reliable highway linking one end of an ocean with the other — a place where low-frequency sound travels at a constant slow speed, while keeping the signal intact for many thousands of horizontal miles.

Throughout the Cold War, the U.S. Navy tapped into this bandwidth with a vast underwater listening network composed of hydrophone arrays scattered across the continental shelves and atop transoceanic chains of seamounts. First known as Jezebel, later as SOSUS (sound surveillance system), the ears of this network were linked via cables to onshore processing centers, where sailors and intelligence specialists spent countless hours tracking the movements of Soviet submarines right down to the number of propellers they were carrying and, in some cases, what the crews were talking about onboard.

Capable of detecting acoustic power of less than a watt at a range of several hundred miles, SOSUS has extended our sensory abilities into the most distant realms of the oceans. Nowadays, scientists at the National Oceanic and Atmospheric Administration's Acoustic Monitoring Project in Portland, Oregon, eavesdrop on the SOSUS array, and their website provides the opportunity to listen to some of the decidedly unearthly underwater noises.

But before we can actually detect many of these sounds, they need to be sped up to within our hearing range. When that happens, we can hear what sounds like a distant truck traveling down a deserted road on a rainy night — actually the cacophony of an undersea temblor somewhere near Japan. A sound of the turbines on a helicopter winding down for seven minutes, which was recorded on three widely separated monitors spanning twelve hundred miles of ocean, has never been identified and never heard again. A strange, circular whooping noise like a police siren running out of energy has been linked to an undersea volcano in the South Pacific previously thought inactive.

One intriguing noise resembles a *Twenty-thousand-Leagues-Under-the-Sea* soundtrack: a low glugging burp, as if enormous air bubbles are escaping an underwater jug. Called Bloop, it bears the distinctive and varying frequency signature of a marine creature, yet is far more powerful than any other known biological call — louder even than the 188-decibel whistle of a blue whale, which is louder than a passing jet. Spanning three thousand horizontal miles of ocean, farther even than the reach of the great whales, Bloop may be the work of the elusive giant squid.

Researchers also recorded a sound in French Polynesia that may have been generated by a huge, fifteen-hundred-square-mile iceberg known as B-15B drifting in the Ross Sea near Antarctica. Sending out a deep, pure sound with a frequency of just a few hertz, B-15B rang like a bell — perhaps a result of the berg cracking or spouting water or scraping the ocean floor. Some scientists theorize that Antarctica and its melting ice sheets may be important megasound generators in the oceans.

In the few years it has been accessible to researchers, the SOSUS array has helped solve perplexing puzzles, including how it is that the great whales navigate the ocean basins. Scientists at the Acoustic Monitoring Project have learned to distinguish whale calls, and thus to track individuals on their travels through the North Atlantic and the eastern North Pacific. Christopher Clark, a biological acoustician at Cornell University, has used these data to superimpose a SOSUS-made track of whales onto maps of the ocean floor. His work reveals how the behemoths skip between underwater mountains hundreds of miles apart. Projecting their calls off geologic structures many leagues away, the whales apparently listen for the return echoes to guide them — as if we could yodel our way between a peak in the Sierra Nevada and one in the Rockies, blindfolded.

Meanwhile, Mary Anne Daher and others at the Woods Hole Oceanographic Institute are tracking an unknown loner with a fifty-two-hertz voice unlike any other species of baleen whale (most sing at between fifteen and twenty hertz). Traveling the North Pacific, this mysterious presence covers routes that bear no semblance to known whale migrations. After listening in for thirteen years, all the scientists know for sure is that this traveler's voice is growing deeper and raspier with age, just as our voices tend to do.

9

Big Songs

HUMPBACK WHALES (*Megaptera novaeangliae*) arrive in the waters around French Polynesia during the austral winter and spring (July to November). Although little is known of their migratory route from the krill-rich waters of Antarctica, it's likely they navigate using the underwater topography of the South Pacific. Perhaps they pathfind using notable, persistent icebergs, such as B-15B, then slalom between islands, distinctive reefs, and seamounts until they arrive at the jumbled underwater heights of the Tuamotu Archipelago.

Supporting this hypothesis are a few photo identifications. Individual humpbacks from French Polynesia occasionally appear in photographs in the Cook Islands, and some from the Cooks are seen in Tonga, while some from Tonga appear in New Caledonia, and a few from New Caledonia appear in New Zealand. Scientists theorize that the humpbacks of Oceania behave much like the peoples of Oceania, maintaining unique cultural identities within island groups, while sharing cultural exchanges across vast ocean distances.

The ancient people of Polynesia likewise traveled these routes in the course of colonizing the islands of Oceania, and they may well have tagged along on the journeys of the ocean's wildlife in order to facilitate their own travels toward familiar places and toward new frontiers. A Maori story tells how the small seabird *ku-*

aka (bar-tailed godwit) guided people from Alaska to Aotearoa (New Zealand) along its eight-thousand-mile migratory route. The story of Paikea, the mythic ancestor of the Ngati Porou people of Aotearoa, describes how he left his homeland of Hawaiki and traveled to New Zealand by riding on the back of a *taniwha* or whale (the same legend that inspired Witi Ihimaera's novel *Whale Rider*). Some Maoris say the inspiration for the traditional double-hulled outrigger *waka* (*va'a* in Tahitian, *wa'a* in Hawaiian) canoe arose from observing the way pairs of whales sometimes forge side by side in rough seas, enabling the smaller and more vulnerable members of the pod to travel in the smoother water of their wake.

However they get here, a few humpbacks arrive in the waters of the Tuamotus each winter, and while the females are busy birthing and tending their calves, the males sing — initially the same song they were singing when they departed the tropics for the trip to the Antarctic the year before. But over the coming season, this old song evolves subtly and continuously, with all the males incorporating the novel phrases until, by October, a new twenty- to thirty-minute opera has been created. Roger Payne, the father of humpback song research, theorizes that male whales are displaying feats not of strength but of memory. Whatever they really mean, these tunes pass from one subgroup to another, appearing on the charts in New Caledonia and Tonga a year or two after debuting in Australia.

I first heard the Tuamotuan humpback song on a mosquito-infested night at the Hôtel Kia Ora, when I abandoned my bed and retreated to the starlight on the dock to fish with a hydrophone and headphones. From the cocktail of chatter below — the snapping shrimp, the territorial calls of fish, the rasping of mollusks, the popping of feeding planktivores — I heard the plaintive song of a humpback whale like a distant cellist practicing in the dark.

At other times and places I have been closer and heard the song of the humpback more clearly, but this singer, suspended upside down in the black ocean (as humpbacks tend to do when singing), bathed in his own reverberations, provided me something better: the eavesdropping excitement of tuning in to a shortwave radio. By adjusting the frequency (the directional hydrophone) and playing with the volume, I managed to catch fragments of this truly foreign language, despite the static of distance and snapping shrimp.

Layered with low-frequency moans, punctuated with whups, yups, yeees, whoos, moos, chirps, and blats, the song seemed to come from another universe. Yet it felt intensely familiar, as it had since I first heard Roger Payne's seminal recordings of humpback songs in the 1970s. These strange underwater voices resonated as archetypal in the full Jungian sense of the word: as familiar as a memory and speaking a metaphorical language understandable no matter what your native tongue. Like the howlings of wolves and the trumpeting of elephants, there is meaning in this music.

If the earth's (or ocean's) hum is the collective voice of our planet, then the humpback is the soloist, Homeric in his scope, indefatigably and poetically in love with repetition and rhyme. Researchers recorded one humpback singing nonstop for twenty-four hours and still singing when they sailed away. Built for travel, the humpback's song is emitted within the forty- to three-thousand-hertz band at one hundred seventy decibels of amplitude — powerful enough to punch through the near-field/far-field boundary. You can see the effects of this if you find yourself close to a singer, as the surface of the water dances from the sheer volume of the song, water droplets bouncing off the agitated surface. Even better, if you are lucky enough to find yourself in the water in proximity to a singer, you will feel the bone-rumbling energy of its voice.

No one yet knows exactly how humpbacks produce their songs in the absence of vocal cords, although evidence suggests they manipulate air through a series of valves and muscles inside blind sacs in their respiratory systems, something akin to a bagpipe. If you happen to be in the water alongside the reefs of a volcanic/coral island when a humpback is singing, you may have the opportunity to observe how the topography of the seascape is used to the singer's advantage. Like a human voice amplified in the shower, where the closed confines and running water amplify the voice, so the already-monumental voice of the humpback is made into something truly titanic by the bowl of the living reef.

Imagine what it must be like for the denizens of the reef who *are* the shower, when after more than a half year of silence, the sea suddenly explodes with the winter's greatest hit. The fish, the invertebrates, the algae, even the waves throb to the beat.

Filming outside a Polynesian barrier reef on a windless day, we come across a pod of about thirty pilot whales, genus *Globicephala* (Latin *globus*, globe; Greek *kephale*, head) — a species of small whale or very large dolphin depending on your point of view. Their heads are bobbing clear of the surface, and the large melons of their foreheads look for all the world like floating bowling balls. The pilot whales are asleep, dangling vertically from the surface while keeping their blowholes dry. So deep is their repose that we are able to cut the outboard, coast up, and slip into the water on snorkel, unobserved.

Gigantic males, slender females, and calves of every size hang from the clothesline of the surface, swaying slightly in the current, as if a collection of large black limp whale-skins are hanging out to dry. Then, completely inaudibly to us, something is heard, something is said in response, and, without a moment of hesitation or a murmur of dissent, all the whales slip back into their whale suits, roll over, throw their flukes into the air, and sound.

That they are running from us is not unique, as this species is normally wary of and even potentially aggressive toward divers. But the whales don't appear to be running away from us so much as sprinting toward something. Instead of sounding to a depth beyond our sight, they remain at about thirty feet, traveling so fast that we can actually see their muscles rippling as sheets of skin slough off like discarded black cellophane, an adaptation designed to decrease their drag in the water.

Intrigued, we track the pod by boat for several miles along the reef until they lead us farther offshore to the spectacle of a pair of wrestling humpback whales. Even from the surface we can see them sparring in tai chi motion, twisting and turning around each other, thrusting with the long swords of their pectoral fins, huge flukes slashing in a scene of unbelievable power and grace.

Slipping underwater, we enter the thick soup of their sound: the groans, the dungeon-door creakings, the roars, the bulletlike *rat-a-tat-tats*. This soundtrack is further layered with the whistling and click-trains of a school of about eighty rough-toothed dolphins (*Steno bredanensis*), who are buzzing like hyperactive mosquitoes around the whales — darting between the barrel-bodies, under the fins, around the flukes. Meanwhile, our guides, the pilot whales, have taken on the demeanor of kids at a carnival, shooting maniacally in all directions, adding their own layer of sound in the form of whistles, squeals, screeches, pulse calls, and echolocational clicks. They join what appear to be twenty other pilot whales.

The sparring whales break off abruptly when a female humpback swims by. She is moving fast, and her exit from our sight is like a gigantic eraser sweeping across the blueboard of the sea, wiping away all three big whales, fifty or so smaller whales, and roughly eighty dolphins. Scrambling back to the boat, we take off in pursuit, following the female humpback, who is sprinting greyhoundlike in the lead, powering just below the surface, visible by the white edges of her pectoral fins and the broad gray beam of her back. When she rises to breathe, the water mounds ahead of her, a shiny blue pressure wave that thins as she ascends and explodes into the spray of her exhale. The tiny bodies of a dozen or so rough-toothed dolphins jockey for position in her bow wave, popping from below to grab a breath, as buoyant as corks in champagne. Escorting her along her flanks are the sinuous black shadows of pilot whales.

Not far behind, the two male humpbacks are running shoulder to shoulder, trying to outmaneuver each other like tandem U-boats. They are also surrounded by dolphins and pilot whales, and, despite the sound of our outboard motor, we can clearly hear the excited, high-pitched whistling of dolphins, the gunfirelike pops of the pilot whales' exhales, and the low whistles of the humpbacks' inhales as the air soughs across the open bottles of their blowholes.

For an entire afternoon we run with them, motoring far ahead, cutting the motor, and dropping into the water on an intercept course. Time and again they pass by us, just a few feet above the thick, cropped cover of the coral reef, the whales never breaking stride, only rolling slightly onto their sides so that we can clearly see their big eyes orbiting in their sockets as they seek to take in the strange forms of us. Unlike some whales in the northern hemisphere who have grown accustomed to snorkelers and divers, these southern-ocean giants have probably never seen human beings underwater, and if not for the important business at hand, would probably stop to investigate, as cetaceans often do.

But this is a drama of life and death, whale style. Not that the males would likely fight to the death (though they might), and not that the female is likely in any danger (though she might be). This is more the drama of evolutionary life and death, with a pair of males vying for access to a female and the chance to pass their genetic legacy forward. The female, meantime, is awarding that privilege, inadvertently or not, to the best whale.

What the pilot whales and rough-toothed dolphins are doing here might be less clear to us had we been observing only from the surface. But from underwater it's impossible to avoid the palpable sense of excitement akin to a crowd gathering at the fight scene of colossi. While the sight is compelling, the sound of the behemoths in battle is like primal screams echoing through a surround-sound theater. Add to this a female humpback, willing or not, and the scene takes on all the rowdiness of an IMAX porn show.

Clearly this is fun on an order of magnitude worth waking from deep sleep for (the pilot whales) and worth abandoning normal daytime rest for (the rough-toothed dolphins) — a fact borne out by subsequent encounters over the years with sparring and/or mating humpbacks in French Polynesia, who are invariably joined by rubberneckers and eavesdroppers of both the cetacean and the hominid variety.

10

The King of
Lake Vaihiria

A FEMALE HUMPBACK with a newborn calf who wishes to escape the attentions of singing and fighting males slips through Avatoru Pass into Rangiroa's ample, and otherwise whaleless, lagoon. Lolling and playing in plain sight of Avatoru Village and its surrounds, the calf rides onto its sleeping mother's back and practices a series of charming, baby half-breaches. The whales are doing what whales have done since long before people arrived in these islands, and the local Rangiroans, although interested, are not concerned. The European expatriates, on the other hand, are convincing themselves that help is in order, and that some kind of a rescue is needed to guide the hapless mother and her infant through one or another of the atoll's passes into the open sea. *La Dépêche,* the French-language Tahitian newspaper, is publishing articles reflecting the general sense of bemusement tinged with wonder at the wayward whales.

I too feel a sense of bemusement tinged with wonder, but mostly at the assumption that whales, who twice a year manage to get themselves from the Antarctic to the Tuamotus along a path that extends thousands of miles, could be silly enough to get themselves lost in Rangiroa's lagoon. This episode is reminiscent of San Francisco's own Humphrey the Wayward Whale, who got "lost" in the bay in 1985, requiring a "rescue," whereupon he returned to the bay, prompting another rescue in 1990.

Whales roam, as people do, and at least some are compelled to explore beyond their known boundaries. This wanderlust is the primary mechanism that enabled humpback whales to populate virtually all of the world ocean. A hundred years ago, a robust population of some one hundred thousand whales wandered the southern hemisphere alone. But with less than 10 percent of that population alive today, humpbacks are largely a novelty and an enigma to the present-day residents of Rangiroa, who gather now to contemplate what measures they might employ to rescue the mother and her calf from their restful shelter in the lagoon.

The Tibetan word *lhun drub* means "spontaneous perfection." As the lama Tenzin Wangyal Rinpoche says, "Everything is just as it is, spontaneously arising from the base as a perfect manifestation of emptiness and clarity." Our experiences of things as less than perfect are based on the dualistic illusion that we are separate from everything else in the universe. In this way the Europeans on Rangiroa are reacting to the novel presence of a humpback whale and her calf in the lagoon with a potent mixture of anxiety, followed by altruism, stirred up by the presupposition that the whales are experiencing fear and confusion, whereas perhaps none is true. Despite my best efforts, I find my own emotions beginning to orbit theirs, and I retreat for a snorkel in the lagoon, hoping to wash away my illusory worry about what the anxious altruists might do to the fearful confused.

Paddling above the coral heads in the water off the Hôtel Kia Ora, I discern the low notes of a distant singer, perhaps the very male the mother humpback is hiding from in the lagoon's shallow water. (Or perhaps not.) From this distance, and at such a low frequency, it is less a sound than an itchy vibration in my bones and teeth. The needlefish I am tethered to by the invisible buffer zone of their lateral lines seem unconcerned with the sounds of the faraway whale, while a giant moray eel (*Gymnothorax javanicus*) bobs and weaves in time to it.

The eel is inhabiting a mostly broken collection of ancient

Acropora coral, bedecked with living sponges, barnacles, Christmas tree worms, and an anemone. At about seven feet long and perhaps one hundred pounds, the eel is a formidable presence in the vicinity of these petite and scattered coral heads. As is her custom, she tucks the bulk of her body into the labyrinth of the fallen and broken coral, with only her head facing into the current, mouth agape and pumping, revealing a pair of wicked fangs as she forcibly breathes in through her mouth and out the coverless holes of her gills.

Despite her fearsome appearance, I know her to be a gentle and inquisitive presence, and so I stop, pleased to see her, an old friend from many previous visits. Beyond the fact that she is very large (most fishes continue to grow throughout their life spans), she is not yet at the uppermost known limit of her species: ten feet. Yet she is big enough for me to know that she is old (close relatives, *Gymnothorax mordax,* the California moray, are still alive at the Scripps Institute after more than forty years in captivity). She is probably decades old and conceivably approaching the century mark.

Her great size also indicates that this is most likely a female, since members of her species are believed to begin life as males and change to females as they age. Like most venerable members of her kind, she has lost the black leopard-spotting of her youth and assumed a plain brown guise, almost monklike in its simplicity. Consistent with her great age, she has also grown impressively heavyset.

The needlefish drift on, but I stay, observing the eel bobbing and weaving with the low-frequency percussion of the singing whale. I match my breathing to hers, allowing my body to sway, my legs and arms to go limp, which causes me to drift off in the current. Sculling back, I continue the synchronized breathing, but this time holding on to a piece of dead coral and adding a dose of *pratyahara,* trying to feel what she is feeling. I can imagine that she is swaying with the whale song, perhaps disappearing into the motion of the water in a kind of auditory disguise. Or perhaps she is swaying because eels are wont to sway. Both could be true. Neither could be true. Further evidence of the perfection of *lhun drub.*

At sunset, showered and dried and now sweaty again, I walk to the end of the dock at the hotel to watch the colors bloom in the tropical sky. Sunsets here are a treat in the way that a trip back in time would be. The northern hemisphere's atmosphere, compromised by pollutants of every variety, reveals impressively corrupted sunsets. Here, south of the equator and far from any continental landmasses, the sky unveils with virgin clarity the anvils of distant thunderstorms, the bulks of whose bodies hide from view two hundred miles away below the curvature of the earth. In this crystalline world, colors are as sharp as knives, slicing cleanly through one another's edges and drawing blood in the western sky at sunset.

As with the coral reef below, this moment of changeover between the daytime and the nighttime shifts is a busy one. Flocks of seabirds stream to and from the open ocean, some going to fish by night, others returning to roost, with fishing boats likewise coming and going, as schools of hunting jackfish ripple the surface.

Most other hotel guests are in the bar at the base of the dock, listening to Tahitian covers of Michael Jackson songs, so I have the end of the dock, this wooden pathway into the lagoon, all to myself. Underscoring the bass beat and castrato voices leaking from the pub are the chattering *crick-crick-crick* calls of the *kirikiri* — terns settling to roost on the rigging of the few yachts that have wandered here on the tradewinds. Coming at me from shore is the raspy *grrrich-grrrich* laughter of the fairy terns, the atoll's ethereal beings with pure white plumage and oversized black eyes, who are tending their single eggs laid on the bare branches of casuarina trees.

The water below the dock ripples, perhaps with the wind, perhaps from schools of browsing fringelip mullet (*Crenimugil crenilabis*), visible through the water. From my position, I can see several schools sorted into age and size classes. The littlest fish are less than six inches long, and their schools are sleek and whippy as cat-o'-nine-tails. They are behaviorally circumspect, skittering only through the protected water around the dock's pilings. In contrast, the schools of the larger mullets, with individuals reaching more than a foot in length, parade aggressively under the dock, around the pilings, and out into the open water, their pectoral fins flipping open and closed like golden fans as they root out the diatomaceous and detrital scum of the sandy bottom. As one of the few species of fish that feed almost exclusively on single-celled plants — diatoms, dinoflagellates, and phytoplankton — vegetarian mullet have a taste unlike any other fish, or so I'm told.

Sucré et huileux (sweet and oily), says the Tahitian woman who has joined me on the dock with her hand-fishing line. She has settled onto the pier not far from me, her bare legs and feet dangling over the edge, the sunset square in front of her, where all sunsets always seem to be in Rangiroa. Like many Tahitian women, her hair reaches past her waist, blue-black and shiny as the inside of a mussel shell, the long plait unfastened at the bottom, the hair holding the shape from habit alone.

She cooks the mullet whole in a pan with just salt, no oil, she says, since the fish are oily enough. Sometimes she adds milk.

I recognize her from the hotel, where she works as a maid, and this dock is apparently one of the perks of her job because I've seen her fishing here most evenings. She seems relaxed and happy at the end of her workday, carrying nothing of the emotional and psychic exhaustion of a Western worker at shift's end — and this is one of the many miracles not only of Polynesia but of French Polynesia, where the Gallic embrace of the sensual allows time for the enjoyment of life as it comes.

We are alternately studying the sky as it fires up in fiery oranges and the mullet below, Heipua dabbling with her fishing hook. The noddies are scrolling across the sky, a Rangiroa-style teletype. The ripples on the lagoon are blossoming and disappearing from the surface, sparking glints of amber and magenta from the sky, a slow-motion aquatic pyrotechnics show. Infinitesimally, the day is melting away and a sultry tropical night is being born. All is still and almost somnolent when we are startled by the sight of the giant moray eel free-swimming through the pilings.

Big *puhi*, giggles Heipua. He will scare all the fish into hiding. Nevertheless, she keeps her hook in the water.

Damn *puhi*, she says, wiggling her line as if hoping to catch her.

She confirms that she knows this old eel well. This is Vaihiria, she says, named after the story.

I am not familiar with the story and am on the verge of asking about it when her two small children — a roughly five-year-old boy and a seven-year-old girl — run down the dock, screaming in excited Tahitian and pointing out to the lagoon. Highlighted against the cherry sky are the unmistakable white cotton-ball blooms of whale blows, motionless in the still air before leaching away in the moist atmosphere. The children can't get undressed fast enough, stripping down to their cotton underwear and flinging themselves off the six-foot height of the dock into the water below, giant eels and humpback whales notwithstanding. Heipua continues fishing, shouting encouragement to her children, who are striking out overarm for the distant whales.

It could hardly be a more different reaction than what I would expect from an American parent. The children are clearly not going to get as far out in the lagoon as the whales, despite the fact that they are as supple as a pair of otter pups in the water. Heipua is not frightened or worried and is — as best as I can tell — teasing them good-naturedly to outswim each other. The kids are buoyant and powerful and make good headway before turning back toward the dock and ascending its rickety ladder to commence a series of cannonballs and somersaults — guaranteeing that Heipua will not catch any mullet today. No one seems to mind, not even Vaihiria, who is sidewinding lazily in the direction of the overwater bar with its glass trapdoor in the floor. In short order, the hotel guests will, for entertainment's sake, drop stale bread through this trapdoor to attract schools of sweet-and-oily-tasting mullet — Vaihiria's favorite, says Heipua.

The old story, told in Tahiti, the Tuamotus, and, in one form or another, throughout the Polynesian world from Hawaii to Aotearoa, is the story of the king of Lake Vaihiria and a beautiful princess named Hina. Although promised to him in marriage, Hina has never met her betrothed, and she sets off for Lake Vaihiria only to discover the horrible truth that the king is an ugly eel. Terrified at the prospect of such a husband, she flees to the protection of the demigod Maui, who has already proven his power by snaring the sun in a flaxen net, clocking it on the head with a fishhook, and forcing it to travel more slowly each day so his mother can have enough time to dry her laundry.

When approached by Hina, Maui not only offers her shelter, but also catches the jilted king of Lake Vaihiria with his fishhook (the one made from the jawbone of an ancestress), kills him, cuts off his head, wraps it in *tapa* (bark) cloth, and awards it to Hina with strict instructions not to lay it on the ground until she arrives home.

Of course she disobeys in that nightmarish way that humans of old invariably do — waylaid by her desire to go for a swim. When she returns to the *tapa* package she sees that it has rooted in her absence and that shoots are springing from it. Waiting beside it, she witnesses the genesis of the first coconut tree, which looks something like an eel head as it sprouts. It quickly grows to maturity, bestowing the first coconuts, the first coconut fiber, the first palm fronds, and all the other invaluable components of *Cocos nucifera* upon her (we can only wonder what gifts she would have received if she had made it home).

According to some versions, Hina's close encounter with marriage then propels her in the opposite direction, into a life of voyaging aboard a *va'a* named *Te-apori*, beside her brother, Ru. Together they discover many islands, including Tahiti-iti (the smaller part of Tahiti island) and Aotearoa. Eventually Hina settles down to making *tapa* cloth, but is tempted one night by the sight of the moon resting islandlike on the horizon. Rowing to it, she steps ashore, abandoning *Te-apori* to the waves forever. Unable to go back, yet concerned for her beloved brother, Hina becomes Hina-Nui-Te-Araara, or Great Hina the Watchwoman, the patron saint of all oceangoing adventurers. You can see for yourself the evidence of her existence in the face of the moon — in the shadows cast by the banyan trees grown from the branches she brought along on her last and greatest seagoing voyage.

This evening, with the moon shining on Hôtel Kia Ora, Vaihiria is no longer headless, but is a large and robust, probably female moray eel visiting the bar. She does this in the night, when the floodlight strobes the shallow lagoon below the trapdoor and the hotel guests — in this case, an Italian couple dressed in identical motorcycle outfits as flamboyant as theatrical costumes, with rolled-up T-shirt sleeves, bandanna headbands, leather vests, and biker boots — are dropping stale bread into the water below. The eel is sidewinding gracefully along the edge of darkness below the trapdoor, where the light is whisper soft, and where her own brown color nearly disappears into the shadows. Despite her disguise, the mullet know she is here, and are bunching up as tightly as an M. C. Escher painting in motion. Yet they are not alarmed enough to abandon this windfall of food.

The moon is tossing silver confetti onto the lagoon. Our Tahitian *mahu* (third sex or transvestite) cocktail waitress is singing quietly along with Madonna's soundtrack from *Evita*. Leaning against the door frame, hip cocked, tray raised, looking hopefully outward, she is bored with Rangiroa and the Hôtel Kia Ora. The hunting and chasing and dying of things underwater, as heard in the thrashings at the surface, and seen in the violent explosions of bioluminescence on the face of the lagoon, are of no interest to her. She is dreaming of a handsome man, a prince among men, and a place where the nightlife is lively. She would trade her world for my world or the Italians' faux-motorcycle world in a heartbeat.

Our tiny tableau holds: the eel swimming in the shadows, the mullet conjoined and skittish, the costumed tourists leaning over the open trapdoor, our waitress waiting, lovely in her red *pareo* and black pearl earrings. It occurs to me that this moment is straight from the heart of *lhun drub*, as pretty and multilayered as the veneers of nacre composing the *mahu*'s pearls.

Everything remains so until a young, tattooed Tahitian man strolls up the pier between the dining room and the bar, and the bubble of our world is burst by his smile, which triggers our singing waitress to dance with a tiny, coquettish shiver of her hips. The man belly laughs and executes a flash of *pa'oti*, the traditional male response to the female's dance: a scissoring of the knees. The Italians look up from their bread duties and — as if waiting for just such a moment of inattention — Vaihiria strikes. The water below the trapdoor geysers, a few drops wetting the floor and the Italians' feet. The schools of mullet explode as if from a cannon. Air bubbles erupt from the acceleration of both the offense and the defense, transforming the floodlit water into a Jacuzzi, and, as usual, everything happens at fish superspeed so that it's impossible to discern the outcome.

But Vaihiria does not return, leading me to imagine that she found success in that moment when the tension holding our tableau together burst. I picture her slithering back to her *Acropora* home, raveling herself in among the sponges, the barnacles, the Christmas tree worms, and the anemone — from where she may choose not to hunt again for days. If lucky, I will see her there tomorrow, swaying to the sound of whale song, or the absence of whale song, or perhaps the whale song that she can hear and I cannot.

11

Impermanence

As mysteriously as they arrived, and without any help from the altruistic expatriates, the mother and calf humpback leave Rangiroa's lagoon. Heipua, fishing beside me during another sunset, tells me they have gone out to where the big fish live, because, obviously, big whales cannot live on little mullet, and a mother whale nursing a baby must be very hungry all the time. I do not tell her that for the most part, to the best of our knowledge, these whales don't feed on their summer breeding grounds. I don't tell her because this may be true, but it may not be true, and I have on several occasions seen various species of baleen whales feeding on their nonfeeding grounds. And, at any rate, I want to hear what Heipua thinks, not tell her what I only half-think.

I do tell her though about the eels and the Romans, as best I can in my imperfect and unitensed French. It's impossible for me to know what she gets out of this. Based on my limited knowledge of the past tenses in French, she may think this story is about present-day Italians. At any rate, I try to tell her how the ancient Romans loved eels and kept them captive and decorated them with jewels and fed them troublesome slaves for dinner. But I don't know the word for slave in French, and am forced to use the only equivalent I can think of, which in this case is *bonne*, or maid — which Heipua herself is.

Her large eyes grow larger, and she studies me to see if this is a joke. I rush to clarify, but fear that my attempt to say that these were Romans from long ago may end up in translation as meaning that geriatric Italians of today, of which there are guests at the hotel meeting that description, keep their maids tied up in (don't know the word for chains) anchors, and feed them to moray eels wearing pearls. *Lhun drub* is slipping fast as I imagine that it was just this kind of poor language skills that led to Captain Cook being eaten by the ancient Polynesians.

But Heipua is a modern Tahitian woman, wearing a T-shirt with a silkscreened portrait saying *Alex Trebek: Canadian Hero*. She is studying me with wider-than-wide eyes and I can follow her transition from politeness to uncertainty to disbelief and finally, now, as she laughs heartily, to dismissal. I am a crazy *popa'a* (red lobster). I am making a joke, and Tahitians love a joke as much or more than the next. She praises my French as being very good, very well said. She tells me that she will let all the *bonnes* at the Hôtel Kia Ora know about these zany Italians. Throughout the rest of this sunset she shakes her head and chuckles and slaps her hand on her knee.

And then, as the Buddhists say, impermanence finally arrives. Darkness drops a silent gavel on the day. Heipua gathers up her fishing line, her mullet catch, and her wet children, and pads barefoot down the dock for home. The sunset, tender and ephemeral, is over. Except it's not quite over; only the false finish is over. The long finish is about to begin, as the sun reaches spider arms from the underside of the horizon to the underside of the clouds, and the sky, grown dark so fast in the tropics, grows light again in the unlikely cathode-ray shades of green and twilight purple. This afterlight lingers, shrinking to a vanishing point along the horizon, growing more intense as it darkens and disappears.

Watching a sunset from its beginning until its true end in darkness is one of life's original invitations to meditation — no can-dle, no lotus position necessary. Because it marks the end of a day, the setting evokes subtle memories of the Big Thing we prefer to forget, and the remembering of this is what lies at the heart of meditation. Buddhist monasteries in China remember it daily in this chant:

This day is already done.
Our lives are that much less.

Even in Rangiroa, the tropical paradise we frequent to forget the grind of our daily lives and where it is inevitably leading, the sun sets, darkness falls, beauty is transient, and the biggest of all darknesses looms. Even in Rangiroa, and perhaps more acutely in the midst of this loveliness than anywhere else, tragedy is a player, even as it waits in the wings.

Because for all its seductive invitations to forget, Rangiroa makes calamities too, and some of these are faithfully recorded in the ancient *puta tupuna* (ancestral books) of the Tuamotus. One such catastrophe — most likely a tsunami sometime around 1560 — destroyed all the villages on the western side of the atoll and damaged others throughout the *motu,* villages that echo in today's place names — Tereia, Fenuaroa, Otepipi, Tevaro, Avatoru, and Tiputa. All that is left of most are the dozens of abandoned taro root pits where people once grew their primary food crop, and the coral and lava stone *marae,* or temple sites.

The presumed tsunami that caused this destruction might have come from distant seismic events along the Pacific's volcano-and-earthquake-laden Ring of Fire — perhaps from massive earthquakes in Japan or Alaska. Or it may have been generated closer to home, possibly from one of the neighboring Society Islands of French Polynesia. Perhaps it was generated by an island collapse, a kind of Icarus-effect inherent to shield (shallow-sloping) volcanoes, which invariably grow too big for the strength of their building materials. Subject to the humbling effects of gravity, these volcanic islands spread and sink until one day they split along the seams known as rift zones running from their summits to the sea.

Sometimes the ensuing collapse occurs slowly, with part of the island slipping beneath the waves in the course of unknown time. In others the collapse is unfathomably disastrous, destroying up to a third of an island in one cataclysmic failure. The spectacular and precipitous Na Pali coast of Kauai, in the Hawaiian Islands, is the cliff left in the aftermath of the collapse of the island's northwestern flank.

Not only do these collapses destroy the host island, they also have the potential to destroy islands at a distance, as thousands of cubic feet of rock suddenly drop into the sea, triggering tsunamis on an unimaginable scale. The 'Alika Slide that carried away a large portion of the west coast of the Big Island of Hawaii one hundred thousand years ago generated a tsunami that stripped soil and rock to a height of eight hundred feet off the island of Kaho'olawe, eighty miles away, and deposited huge chunks of coral reef limestone one thousand seventy feet up the slopes of Lanai, one hundred twenty miles away. In fact the entire island of Lanai is believed to be the remnant of a much larger island that geologists surmise lost its summit and two-thirds of its original mass from an island collapse. Evidence of that disaster is scattered across a seventy-mile-long debris field underwater.

Likewise, the enormous, dinosaurlike coral clasts frozen on Rangiroa's northwesternmost *motu* may have been the work of an island collapse and its ensuing tsunami, perhaps the same collapse that wiped out the villages on the western *motu* around 1560. If so, then even as the villages and temples of the people were being stripped to their stone foundations by the waves, new temples were being built from the bones of the coral reef — the four-million-pound boulders evoking an awe every ounce as deep and resounding as the *marae* of old must have been.

The mother and calf humpback whale return to Rangiroa's lagoon one morning, flaunting their navigational talents and their self-sufficiency in plain view of the village of Avatoru. Their blows sprout and disperse like dandelion spores from a turquoise field. There is no talk now of escorting them back to sea.

Heipua is not surprised to see them again, even though they are the first whales to be sighted in Rangiroa's lagoon in recent memory. Of course they're resting in the lagoon now that they've eaten the good stuff out there, she says, the *aahi* tuna and the *mahimahi*. They've come back here to sleep. It's a good life. *Maitai.*

At noon the next day, when it's too hot and the glare too brutal to film anything topside, and the *mascaret* running too hard to work underwater, I snorkel in the lagoon, looking vaguely for whales, and finding Vaihiria coiled inside a holed and sunken dinghy so encrusted with corals that it's nearly impossible to see the boat anymore. As always, she is swaying and waving, mouth agape and pumping, fangs showing. But this time a pair of cleaner shrimp and a juvenile emperor angelfish are attending to her, one of the shrimp working so far in her mouth that I only occasionally see a flicker of movement, the other flossing her teeth. The juvenile emperor angelfish tending her gills is wearing dark blue with striking concentric rings in white and pale blue, as pretty as a mobile bull's-eye. There are no whales that I can hear, yet Vaihiria is swaying, perhaps because it's all *maitai*. The *lhun drub* of being alive.

I leave her there to enjoy the work of her handmaidens. But she has apparently had enough pampering and swims after me, as big, thick, and muscular as an anaconda. It's always startling to be underwater with a free-swimming moray, particularly a large one, and when I turn and find her dogging my heels, I flinch, instinctively pulling in my arms. But this is Vaihiria, an old friend, and I have nothing to fear. We swim together — her following me, no matter what I do to change the leadership role. And so, just as I took a tour with the needlefish, she is apparently taking a tour with me.

We head out to a place I know where the skeletons of old corals have been tumbled and stacked by the actions of waves, perhaps by the big storm waves of the 1982–1983 El Niño, when an unprecedented six tropical cyclones swept through French Polynesia, destroying reefs to a depth of two hundred fifty feet. In the deeper water farther from the beach, we find hunks of broken corals and reef limestone looking like the abandoned building blocks of an ancient construction site. Most are strewn haphazardly across the sandy bottom, but some look to be purposely stacked, as if by human hands (though they are not). They resemble the archaeological remains of *marae* hidden away throughout the Polynesian islands, their stones dislocated by the snakelike roots of banyan and *mapé* trees.

This mysterious Easter Islandish part of Rangiroa's lagoon is rife with nooks and crannies and secret passageways, offering hidey-holes for an astonishing assortment of marine life. Vaihiria appears to know the place well, as one would expect, and dives headfirst into a crack the size of a mailbox, completely disappearing before reemerging from another hole a good ten feet away.

The water in the inner lagoon is turbid enough to cast a scattering layer of sparkle over everything — the lavender buds of baby *Acropora* corals, the green-and-purple tentacles of a large sea anemone, the feathery yellow cups of *Tubastrea* corals, which are hosting and feeding the equally brilliant yellow wentletrap snail. A pair of Bennett's butterflyfish, mated for life, and bearing startling blue eyespots, kohl-rimmed in black, on their flanks, are drifting side by side through the back alleys of the temple, while bands of roving unicornfish inhabit the midwater, and a school of needlefish hovers at the surface, playing the part of the silver lining in a cloud of ripples.

As a promise of what might follow in the wake of impermanence, this site is truly inspirational, and when I make the long swim back to shore, raising my facemask from the water to check for passing boats, I leave Vaihiria tucked into a kind of domed apse in the reef, sheltering below the limestone remains of a plate coral.

Although I can't possibly know it, this is the last time I will see her. Sometime in the course of the few short years between this visit to Rangiroa and the next, the Hôtel Kia Ora is sold by its

French owners to Japanese owners, who decide that the large, free-swimming eel in their portion of the lagoon is a threat to their snorkeling guests. Concerned with risks in a tropical wilderness they would like to portray as a Disneylandlike paradise, the new owners hire someone to swim out to the coral heads with a speargun one sunny morning or afternoon and shoot the eel. Such is the end of Vaihiria, the queen of Rangiroa's lagoon.

In the Tibetan tradition of Buddhism, six realms of existence, or dimensions of consciousness, prevent us from experiencing the *lhun drub* of the world. In one of these realms greed rules, embodied by the *pretas,* the hungry ghosts — beings with bloated, famished bellies, but only tiny mouths and throats, who inhabit a parched land where water is nonexistent for centuries at a stretch. If the hungry ghosts do find food and drink and get it past their constricted gullets, it bursts into excruciating flame in their stomachs. Suffering thus for all infinity, the *pretas'* endemic greed can never be filled because they are incapable of looking anywhere but outside themselves for sustenance, when the answer lies within: the antidote of generosity.

12

The Infinity Pool

THE NEW OWNERS of the Hôtel Kia Ora have made many supposed improvements since assuming management, and the latest is a swimming pool spilling over an edge that appears to bleed directly into Rangiroa's incomparable lagoon. The cross-hatched geometry of the thousands of cobalt-blue tiles and white grout gives the pool something of the feel of an unfinished 3-D animation project, as if the designer is still fine-tuning the angles on a computer screen. Yet despite the mirage, the pool is real, so real that it makes the world around it less real somehow. And the lagoon, that turquoise gem filled with coral heads and all their attendant pleasures and hazards, becomes little more than a backdrop to the pool: an enormous screen animated with images of seabirds chasing schools of fish, whose panicked underwater skiddings ripple across the surface of the lagoon like invisible feet bunching up a turquoise rug.

All of this unfolds as the hotel's guests lounge, sipping tropical drinks and watching or not watching the real world out there, where things are chasing and fleeing and dying and being eaten. For the most part the guests choose not to watch, I notice, and so the beauties and dangers of this atoll devolve into a visual Musack. Perhaps because of this, the pool is a hit, a place desirously immune from reality, and during this visit to the Kia Ora I never see a guest swimming in the lagoon, despite the fact that most of them are paying more than five hundred dollars a night for an overwater bungalow complete with a private swim step leading into it.

I like the new pool, too. In the fourteen years that I've been coming to this hotel, this is the first time I've been able to return from a day of filming, feeling sunstruck from hours on an open boat, and fling myself into the brackish water to cleanse the sweat and the stale suntan lotion. Someone will bring me a drink in the pool too, which they will not do in the lagoon, as well as little plates of fresh coconut and black olives marinated in mustard.

For centuries the lagoon was the island's obvious relief from the sun and the heat, and it still is for Rangiroa's residents. But for guests at the Hôtel Kia Ora, the pool is the new recreation center, although no one actually recreates in the pool as it's too small to swim in. Instead, during the last hours of daylight, when the lagoon is filled with Tuamotuans, the pool is filled with guests relaxing after their scuba dive or their visit to a pearl farm, glancing up occasionally to check the screen of the lagoon as it delivers its projections of the natural world.

Out on the water, caught up in the excitement of the hunt, a flock of *kirikiri* are acting like blackbirds in the stubbled fields of autumn. Swirling and bunching with elasticized precision, they gather above the lagoon, their dimensions expanding and contracting as the flock lurches between the air and the water. These noddies are about half the size of your common coastal gulls of North America, diminutive, and, like all terns, as pretty as butterflies. But whereas virtually all other terns are bedecked in white feathers with flippety black crests atop their heads, the *kirikiri* are photographic negatives: smoky, sooty-colored beings, with dark plumage blending upward in airbrushed perfection to smooth white foreheads.

To the outside observer, lounging in the infinity pool with a pineapple-and-coconut-milk cocktail, their method of hunting appears to be composed of equal parts daintiness and chaos. Rather than the dramatic bill-first dives that most terns execute, noddies skip close to the surface, occasionally dropping into featherweight belly flops, but more often simply dipping their bills to pluck their prey from the surface. A flock of noddies working the water in this fashion looks like a visual representation of music, the black birds acting the parts of the notes, fluttering up and down the scales of the sky.

Their aerial gavotte is designed to overcome the limitations of finding tiny fish in a huge ocean, since reconnoitering alone might well prove metabolically more expensive than its payoff. In the loose company of others, with hundreds of eyes continuously scanning, the flock becomes a superorganism capable of locating food over a vast distance.

For flocking birds, virtually every aspect of their lives is influenced by the behavior of their flockmates: preening when others preen, sleeping when neighbors sleep, rising from the nest into the air when a flockmate gives an alarm call, initiating courtship when others are courting, too. In a sense, these birds with their much maligned bird brains are part of something bigger than themselves — a connected supermind that takes all the sting and the confusion out of solo decision making.

At their best, when sweeping across the water in tight formation, a flock of *kirikiri* acts like a single loose-jointed creature, stretching and pooling with the fluidity of liquid. In a seminal study of the shorebirds known as dunlins, researchers used high-speed film to decipher some of what goes on inside a flock of birds (film shot at high speed and played back at the usual twenty-four frames per second appears as slow motion). In the extreme slow motion footage of the dunlins, the researchers observed that the rapid changes of direction executed by the flock were not initiated by all the birds simultaneously but started from a single bird, or a few birds together — a movement that then radiated through the flock like a wave. These maneuver waves took only fifteen milliseconds (fifteen-thousandths of a second) to pass from one neighbor to another.

Yet when tested in the laboratory, the dunlins' fastest reaction to the sudden stimulus of a flash of light was only thirty-eight milliseconds — too slow to account for the rapid response observed in individuals within the flock. The researchers concluded that the shorebirds reacted not to their immediate neighbors but to the maneuver wave itself, with individuals anticipating its arrival and changing course accordingly.

But one wrong move, one minute shift in direction between the onset of the maneuver wave and its arrival at the individual, would result in chaos, as the flock self-destructs in a monumental collision of broken wings. For the maneuver wave to persist as a beneficial behavior it must be foolproof. And anyone who has watched the rapidity with which a flock of shorebirds changes direction knows that maneuver waves are not constant events that express themselves outward to the edge of the flock before the next one is initiated. Instead, they morph unpredictably, like a tidal wave in a bowl.

Whatever the mechanism, the scene is the same most evenings. From the comfort of the infinity pool, the lagoon's surface appears glassy — until a school of baby fish bursts through, missile-like. The flock of *kirikiri* reacts as a single organism, cranking on its heels, fluttering down, as a dozen birds dip to the surface and return with fish flashing between the black forceps of their bills. Yet that's hardly the end of it. Within seconds, the black kites of frigatebirds, whom the Tuamotuans call *ota'a*, appear in the sky, wings tipped sideways, spilling air as they descend.

One *ota'a* singles out a *kirikiri* and the two begin a dance, more like a mugging, their wings flipping from one tack to another, sending the noddy skidding away from the frigatebird, who responds by wheeling around and threatening a collision. Brakes are applied. Tail feathers fan out. A second *ota'a* intercepts the signal and sweeps in close enough to pin the noddy to the sky, where, panicked and fearful, it regurgitates the fish it swallowed only seconds ago. From above, a third frigatebird arrives at the falling morsel and scoops it up.

This daily drama is visible and audible from the infinity pool — although it competes for attention with the stereo system in the bar playing "Girls Just Want to Have Fun" in a French-Tahitian combo, and with a small group of Americans splashing in the pool. Drawn together by their inability to speak French, these visitors are as noisy and disruptive as starlings, engaging in an intense bout of social signaling, sorting out who lives where, who does what for a living, who drives what at home.

By eavesdropping, I can learn much about my fellow travelers to the coral world: the American woman from Queens with a painfully sunburnt back and a powerful, nearly operatic voice; the two couples, unknown to each other before this moment, polite midwestern types who would probably not be caught dead back in Wisconsin, or wherever, bragging, but who are nonetheless inspired by distance now, until all five are engaged in a bout of cheerful, competitive spraying. They are comparing their dive adventures: who has better gear, who knows what comprises better gear, who has traveled to more places, who has seen bigger and scarier things while there. They have an obsession with sharks — not surprising in light of the fact that sharks are the signature attraction at Rangiroa. But beyond this obsession, they appear to have no real interest in sharks. Judging from their conversation, their interactions are solely narcissistic: not what the animal did in relation to its natural world, only what it did in regards to them. *It came this close . . . It lunged unbelievably fast . . . It freaked when we came around the corner and found it sleeping . . .*

Meanwhile, out on the lagoon, a juvenile bigeye trevally caught by a noddy falls back to the water to be reingested by the ocean. Sometimes, with a big enough flock of noddies and an armada of aggressive frigatebirds, the little fish fall like rain, their reentry points marking the surface with the bull's-eye targets of ripples.

If you happen to be snorkeling or diving in the vicinity of a feeding frenzy while this is going on, you can see the little fish streaking like lightning bolts for the company of the school. With a homing sense as true as any bird's, they dash toward their beleaguered fellows, gathering until they form a subschool of survivors. If the subschool grows large enough, it may temporarily abandon the mother school, particularly if the battle is not going well for the latter.

Watching, you think it must be hopeless: that every last living creature in the sea or in the air desires these little fish, as bright as newly minted silver coins. Tossed cruelly into the hordes of the greedy, hounded day and night, juvenile trevallies live lives of unending terror under conditions of perpetual warfare. Yet they survive, many of them, somehow.

A school of predatory jacks, which the Tahitians calls *paaihere*, are dogging their heels from below. The noddies are skipping across their spines from above. The frigatebirds lurk in the shadows of clouds. Bunching together, the young fish spill and pool and reunite like loose mercury. When the *paaihere* attack, the little fish perform a defensive maneuver known as the fountain effect: the school instantaneously splits into two as each subschool reverses direction and circles behind their attackers on opposite sides. This behavior awards the trevallies a running headstart from their enemies.

Capitalizing on the momentary advantage, the little fish dive, trying to escape the reach of the birds above. But the *paaihere* are relentless hunters with big appetites and phenomenal accelerating abilities. They wheel around and come at the trevallies from below, herding them toward the sunlight, where the reflections of their bodies sparkle in the irises of the dancing noddies.

Trapped, the little fish unloose another defensive stratagem known as the flash expansion. Without warning or any apparent means of coordination, the school explodes, and members scatter shrapnellike out from an imaginary center. The whole reaction takes place in as little as twenty milliseconds, as each fish accelerates to speeds of twenty body-lengths per second. Amazingly, the trajectory of each trevally carries it away from the center and away from its neighbors and is accomplished without any collisions — which would surely prove fatal under the circumstances. More astounding is that the entirety of the flash expansion occurs at speeds faster than the rate of nerve impulses traveling from the little fishes' eyes to their brains and back to their muscles. This defensive play has nothing to do with sight, and the little fish are as good as blind throughout it.

Curiously, computerized fish programmed to behave like a school of fish cannot perform as well as the real thing unless they are subject to some kind of a unifying field, which is itself influenced by all the individuals in the school, and which in turn links them together. Sight alone is not enough to maintain tight schooling. Neither are the lateral line systems that enable them to sense minute pressure changes in the water. Real fish that have been temporarily blinded with opaque contact lenses, and others that have had the nerves to their lateral lines cut, still manage to school effortlessly. Biochemist Rupert Sheldrake of Cambridge University proposes a controversial solution, suggesting that schools of fish and flocks of birds are coordinated by morphic fields that hold them together, and that both influence the individual and are influenced by the individual. Isolated from their flocks, most flocking birds will make Herculean efforts to reunite. Isolated from their schools, some fish, notably herring, die.

Afloat in the waves, we await the arrival of our Zodiac. We have run through every frame of film shooting silky sharks, silvertip sharks, and oceanic whitetip sharks in the blue water offshore, and are simply riding the current toward the pass. We are happy enough to drift and wait, our BCDs inflated like oversized pufferfish, our cameras out of film, dangling from the leashes on our wrists, though we are also tired and wordless and ready for a drink in the infinity pool.

Yet the end of this day has something more to offer. The birds arrive first, guano and the occasional feather raining down. From our position, we are granted an intimate view of the choreography of the flocks, the elasticized precision, the muggings, the evasions. I tilt my head down and discover a ball of juvenile fish flashing and turning, stretching and collapsing like a a hyperactive screen saver. In the darker water below, the *paaihere* lurk, their engines idling, until suddenly all hell breaks loose at a rate of speed human eyes cannot follow. Everything blurs, as if we have suddenly jumped dimensions. Without knowing what is happening, I instinctively stop paddling and pull my limbs into a floating fetal position.

Then, just as quickly as it started, the blur slows and begins to sort itself out. Little fish are everywhere, confettilike, glinting all over the sky of the sea. The feet of noddies dip and retract through the surface like wary bathers. In an instant the tables have turned and the *paaihere* — befuddled by the flash expansion — are on the run themselves, carving hard U-turns as they drop into the depths or flee for the surface. Powering through from below are the silvertip sharks, the linebackers of the reef. The *paaihere,* so deadly a moment ago, are now little more than hors d'oeuvres.

As if to flaunt this reversal of fortune, the silvertips travel with a contingent of juvenile golden trevallies — close relatives of the juvenile bigeye trevallies. This species spends its yellow-and-black-striped infancy riding the bow waves flowing from the noses of sharks. Too small for the silvertips to concern themselves with, the golden trevallies scrounge leftovers from their hosts' meals — in this case, bits of *paaihere* torn to pieces and floating away.

We take it all in as best we can — disadvantaged by fish superspeed, awed into silence, eyes wide, cameras empty.

13

Inshallah

'M ALONG FOR the ride this day shooting topside camera, a job that enables me to spend a lot of time in the presence of Manu, our Zodiac driver, both of us fending off the dual killers of the austral summer: the sun and the southern hemisphere ozone hole. While we wait for the divers to return to the surface, we take shelter — I under a large blue-and-white-striped umbrella from the Hôtel Kia Ora, Manu under a huge, weatherbeaten beach umbrella, complete with the ubiquitous Hinano (Tahitian beer) logo.

To scan the entirety of the horizon, we sit on opposite sides of the Zodiac's pontoons, looking past each other. One of the great pleasures of being on the sea with nothing much to do is the leisurely pursuit of watching and waiting, like being a fire lookout or a lifeguard, minus the responsibilities. We are confident that something will happen, as it always does on the ocean, although exactly what, when, or where is completely unpredictable. Perhaps it will be a school of flying fish bursting from the still water and soaring away on dragonfly wings. Maybe a dolphin will appear, its whistling call piercing the surface as it turns on its side and lays its curious left eye upon us.

Because I'd like to see one, I ask Manu whether he thinks there might be any marlin. He answers, *Inshallah,* and laughs. His laughter rumbles through his body, all three hundred fifty pounds of it (I'm guessing), rippling up and down his oversized belly and thighs, shaking the pontoons of the Zodiac, shaking me and the umbrellas, before traveling onto the water, where, for all anyone knows, Manu's laughter might well be destined to trigger a distant tsunami. He laughs a lot, and says *inshallah* a lot, a phrase he picked up from an Israeli, no, Arab, no, Palestinian — he's not sure which — film crew he worked with.

Does he know what it means?

Oui, he says. It means if the gods would be so kind as to allow it to be so.

While he's telling me this I see the thing we have been waiting for off in the distance, on the edge of the highway of the *mascaret,* a huge splash — the belly flop of a breaching marlin.

Jesus! I shout, and Manu laughs again as he folds his battered beach umbrella and pulls the ripcord on the outboard. He is not in a hurry, one of the things I imagine you abandon once you cross the three-hundred-pound threshold. The fish are out there. The birds, too. Everything is always there. So why hurry? The sun does arise, and make happy the skies, wrote William Blake.

We motor along at a leisurely rate, feeling the leisurely slap of the waves against the hull, feeling the somnolent breeze stirred up by our five-knot pace. The thrashing below the distant birds is a school of *aahi* tuna, Manu says, which work the edges of the *mascaret.* Where there are *aahi,* there are usually marlin, as well as sharks.

Manu makes his living driving Zodiacs, delivering divers to and from select points on the reef and then performing the really important task of finding them again when they surface in the unpredictable places Rangiroa's powerful currents carry them. But at heart he is a *pêcheur,* a fisherman. I've noticed this with all the boatmen in French Polynesia. They can't be restrained from using the Zodiac to chase fish or to surf waves — their twin Tahitian birthrights.

Manu is indisputably heavy and slow. In a way he has begun to resemble a marine mammal, and when we arrive back at the dock at the end of each day he lowers himself into the shallow water and lolls like a dugong gratefully relieved of gravity. But at this moment, as we mosey up to the feeding frenzy of fish, billfish, birds, and sharks, the look in his eyes is that of a predator. If not for me, I suspect, he would retrieve a stash of hand-held fishing line from under a pontoon and get to work catching himself some supper.

He wants me to understand what we are seeing. The boiling white water is the mark of the *aahi* tuna, who are frantic with the chase. The noddies are dancing near the wave tops, but warily, because the tuna are not averse to snatching little birds as meals. The marlin are following the schools of *aahi,* each marlin awaiting the opportunity to spear a tuna on its bill, jump out of the water, toss it into the air, and swallow it whole, headfirst.

Manu's voice rises as he explains these things, and I realize that this is sex for a fisherman. He points and says things in Tahitian I can't understand, and when he switches to French, his second language, it's hopelessly garbled. It doesn't matter. I know what he's saying even if I can't understand the words. The *aahi* are thrashing a school of *mahimahi,* the slender beauties of the tropical sea, their luminous bodies flashing in blue, gold, green, and purple. Those *mahimahi* not under attack are pursuing their own prey, hunting *marara* (flying fish), who are abandoning the water in droves, scores of them launching through the surface, spreading their pectoral and pelvic fins wide in order to soar off at crazy angles, their long tails whipping through the air and digging into the surface to renew their glide. Because of the refractive differences between the water and the air, their escape paths are hidden from the *mahimahi* below.

But, as we know, the air is not safe. It's filled with frigatebirds, who enjoy dining on flying fish better than anything else and therefore spend a considerable amount of their lives in the company of *mahimahi* — who can be relied upon to chase the flyers into the dry world. Soaring on pterodactyl wings, their necks craned downward at a ninety-degree angle, the frigatebirds drop onto the backs of the flying *marara* and flip them from their flight paths into their own gullets. In a reversal of the situation in the lagoon, little noddies dance beside the frigates, hoping for a dropped fish.

Despite his name (Tahitian *manu:* birds), Manu is not interested in birds unless they tell him something about fish. In this case he's adamant that I see something in particular — something different, he says. He points to another group of frigatebirds, the ones accompanying the maelstrom where the *aahi* are tearing into the *mahimahi*. Blood is streaming into the water, turning the froth pink. Manu explains that not all of the blood belongs to the *mahimahi*. The frigatebirds are divebombing the tuna and pecking at the soft, scaleless surface of their skin, excavating meat and muscle from the wound.

What a rude shock for a tuna, one of the masters of the sea, with a streamlined body as close to a torpedo as any living thing. Tunas are built to kill, and this harassment by birdbrained featherweights must be humiliating (if tuna are capable of humiliation), not to mention painful.

Everything about a tuna is designed for speed. The way their dorsal fins fit into slots in their bodies, the way their eyes lie flush against their heads. Unlike other fish who undulate, tuna project muscle force from the midbody backward, enabling the tail to act as a hydrofoil and maximizing their forty-mile-per-hour bursts of speed. In fact, the capacity for speed is such a defining characteristic of the tuna lifestyle that it has transformed them from ectothermous (cold-blooded) beings into endotherms (warm-blooded).

Whereas heat — the by-product of the functioning of cells and muscles — is held within the bodies of endotherms by insulation (blubber, fat, or feathers), in fish it's shed through the gills. Yet tuna generate such furious heat in the course of their lifestyle that they have developed a novel system: a countercurrent heat exchanger built into their circulatory system, where outward-bound and inward-bound blood vessels lie alongside one another, and the blood warmed by the action of the fish's muscles transfers its heat to the cooler blood flowing inward. Hence, metabolic heat stays at the tunas' core, enabling them to maintain a temperature as much as sixty-eight degrees Fahrenheit higher than the surrounding water. The luxury of perpetually warm muscles increases the tunas' response time, empowering them to roam cold waters while remaining speedy and alert.

Yet here are the frigatebirds pecking at them as if grabbing sashimi on the run. Meanwhile, the blood trailing from their open wounds mixed with the pungent oil of tuna flesh is the most powerful shark attractant on the planet. Manu points out the silky sharks and the silvertip sharks, arriving to chase down the warm-blooded, protein-rich *aahi*.

It's the same in the sea no matter your size. And so the giant tuna adopt a defensive system identical to that used by the juvenile bigeye trevallies in the lagoon: bunching together, moving as one, flashing their colors. Every few moments an attack from a shark sends a subschool bursting into the air, where, without the wings of flying fish, they can only arc across the surface before reentering somewhere new. Manu is laughing again, and waving at a small fishing boat speeding this way. It can't get here fast enough, in his opinion (even though we meandered), because the two fishers aboard are the proxies of his imagination: slim guys still able to partake of fisherman sex and not just ogle from the sidelines.

Energy is eternal delight, wrote William Blake, and we are part of the energy too, even if we are only observing.

14

Poi Dogs

SOME EVENINGS I BYPASS the infinity pool and take up station on one of the hammocks strung between coconut palms outside the waterfront bungalows, where I am not staying, and therefore am perhaps not entitled to sway. I enjoy the hammock, surrendering to the slight queasiness of it with its sedativelike qualities. Just down the beach, families of Tuamotuans are enjoying the lagoon: old women entering the water in their *pareos* (sarongs) to bob blissfully on their weightless haunches, kids thrashing around on boogie boards, young lovers towing each other across the surface.

Rangiroa's fish-eating dogs are at work in the lagoon too, in this case a pair of island-skinny rottweilers who perch tremblingly on the submerged cinder blocks laid down for mooring buoys. The dogs have awakened from their midday comas and are enjoying the cool breezes while gathering the motivation to fish. Their backsides planted underwater, they stare intently at the ripples in the waves.

Their concentration is contagious, and I find myself holding my breath, willing them success in their seemingly impossible task of sorting out which swirls in the water are fish, and which are made by children at play, and, if the swirls are made by fish, which are fish worth catching and eating, and which are fish who might carve you up with their built-in scalpels, or poison you with one of the ocean's many venoms. The dogs are skinny enough that you sense they need to do this work. Yet they are also eager enough that you feel they love this work.

I am delighted to see them. Years ago the island was filled with packs of fishing dogs that worked even the dangerous, surf-swept edges of the outer reef. Moving at a trot, they displayed the curly-tailed, pointy-nosed phenotype of feral dogs all over the world, with the added characteristic of the Rangiroa short-legged variant — as if their not-so-distant ancestors were corgis or basset hounds. But research from the island of Rarotonga, eighteen hundred miles away, has shown that the dwarflike dogs of Polynesia were present long ago, even when Captain Cook first made landfall — dogs his naturalist, George Forster, described as "very low on the legs." The Polynesians call them *poi* dogs.

The roving gangs of *poi* dogs that inhabit Rangiroa today are, like their predecessors in Rarotonga, still occasionally eaten by people, and perhaps because of this they are neither friendly nor unfriendly to my species. Most individuals maintain a loose affinity with some house or houses they might marginally consider home — what anthropologists refer to as community domestication — yet it's clear that their real allegiance is pledged to the nation of dogs. People toss them fish tails or pork ribs from their own suppers, but not enough to live on, and so the dogs of Rangiroa, as elsewhere in the Tuamotu Archipelago and the South Seas, teach themselves to fish.

The purebred rottweilers working the edge of the lagoon are obviously first-generation Rangiroans, yet they are already as gimpy and battle-scarred as their feral brethren, having clearly taken to the fishing business with zeal, learning from observation of their fellows, as well as from their own trial and error. Each evening the two perch twenty feet apart on the separate, submerged cinder blocks, where a promising ripple causes them to lean forward, ears cocked, bodies vibrating with excitement.

Most of these ripples pass by to break on the beach without the dogs acting on them. But every twenty or so wavelets, one of the rottweilers launches itself into the water, dog-paddling furiously, and whimpering with anticipation. In about four out of five launches, the attack peters out and the dog simply carves a wide circle and paddles back to the perch. The other 20 percent of the time, the dog commences the second phase of the attack, dipping her head underwater and transforming herself from a dog-paddler to a full-fledged diver.

From the hammock I can sometimes see what they're chasing. A baby blacktip reef shark, about eighteen inches long and as floppy as its cartilaginous bones would imply, attracts the dogs' repeated attention. But none of their paddling attacks bring them close before the shark shoots off with a flick of the tail. Sometimes the rottweilers are fooled into thinking that tight schools of small fish are actually one large fish, which they pursue until their prey shatters and flits away. Sometimes they launch separate attacks on the same butterflyfish, coordinating their work, driving the fish toward the other dog, eyes darting frantically from their prey underwater to their ally at the surface. This happens so repeatedly and successfully that I realize it's part of their design, and the reason they work about twenty feet apart in the first place, utilizing ancient skills honed on herding ungulates.

Their catches are made in a flurry of underwater splashes, hind legs clawing at the surface, rumps bobbing like buoys. If successful, they row their prize to shore in a snout held high, eyes half-closed, as the fish's tail whips around. No matter the size of the catch, the rottweilers share with a delicate formality, lips grinning and bristling with fish bones.

The only time they bypass this confederation is when food is abundant, as I see one evening when a fishing skiff pulls ashore, laden to the gunwales with a breeding school of humpback snappers caught in a fish trap in Tiputa Pass.

When I stroll over to have a look, I find the fishers aboard, two men and one woman, uncharacteristically hostile. (Of late there has been talk in Rangiroa of limiting fish catches, at least in the fish breeding season, and many fishers don't like it.)

When I ask these Rangiroans if the pair of large, mustache triggerfish, pathetically gasping for air in the bottom of the skiff, will be their supper tonight, the woman warms up a little, smiling and telling me they are the best eating of all. But ten minutes later the fishers have lobbed the nearly dead triggerfish into the lagoon. Not enough fat on them, the woman says, too tough to eat.

The rottweilers are unconcerned with such details, however, and charge after the twitching carcasses, where each does battle with the equally hungry baby blacktip reef shark. The bits and pieces that each dog manages to wrest ashore it eats by itself, a rare double catch, while the human fishers unload a short ton of snapper into coolers destined for the last evening flight out of Rangiroa, for the restaurants of Pape'ete. For all the diners unaware or unconcerned with the true cost of their meals.

I am delighted to see the fishing dogs of Rangiroa because every decade or two the islands convulse under a dog pogrom. In 1983 a purge swept through the Tuamotus after a pack of dogs entered a house on the atoll of Manihi and ate an unattended baby. Islanders reacted with guns, knives, poisons, harpoons, whatever. Those who wanted to save their dogs hid them and barred entry to their houses. No one knows how many dogs died, but enough lived to breed the population back until a happy equilibrium was reached, and once again dog packs patrolled the tide pools and beaches.

Then, a few years ago, in reaction to a couple of nervous tourists, the Hôtel Kia Ora implemented its own purge of dogs and cats on Rangiroa. Any dog on the loose — which is to say every dog on the island — was fair game, and hundreds died. Now the island seems empty of dogs (and cats). The packs that used to wander the hotel compound are gone, and during my stay I see only one dog on one misty afternoon, gleefully raising his leg on every unclaimed palm tree, bungalow, golf cart, and luggage dolly on the grounds. His joy at being off the map in a canine terra nullius of pissingly good proportions is obvious, and I find myself secretly cheering him on — then fearful when a hotel gardener strolls his way. But apparently the islanders have nothing against dogs at this moment, and man and dog pass with little more than a glance.

This antidog philosophy is only one example of how the Hôtel Kia Ora has changed in the time I've been visiting. Fourteen years ago, under French ownership, the hotel had the typical laissez-faire attitude of businesses run by temperate-zone Europeans relocated to the tropics. Dogs were allowed to roam at will, enabling the guests to observe their fish-catching skills on a regular basis. In equal numbers, long-legged cats clawed their way up the hotel's hundreds of coconut palms — the last vestiges of the land's former incarnation as a copra plantation. Adding to the authenticity of island dogs and cats and all their fleas were the palm trees that loosed percussive rounds of coconut ammunition with unnerving regularity.

During my first visit to the Hôtel Kia Ora, a guest was beaned by a coconut and subsequently flown to a hospital in Pape'ete. Yet the fruit were left hanging ominously in place for many more years afterward because the French management believed such hazards were part of the adventure of the Tuamotus. Now, however, the *petite* adrenaline thrill afforded on the walk between your bungalow and the dining room has been neutralized, as the Japanese owners trim not only the palms around the hotel, but also the palms on all the acres of surrounding grounds, once as wild and challenging as a live-fire artillery range.

Conscientiously and unimaginatively, the new owners have stripped away much of the gamble of this place. I miss the old bungalows, absent of door locks, air conditioning, or fans. Although the rooms were stocked with mosquito coils, the mosquitoes thrived on guest blood. Now, at the behest of the hotel, the entire island is sprayed with pesticides twice a year, and the hotel sprays its own grounds every month.

Sadly, what's comfortable for the guests is decidedly uncomfortable for the reefs that attract the guests. Free of pollutants since their inception millions of years ago, the tiny coral animals in the lagoon and on the outer reef slope must now adjust to the presence of compounds designed to kill tiny forms of animal life. Unwittingly, the new ownership of the Hôtel Kia Ora has launched an entirely new kind of adventure.

From the hammock one hot, muggy evening, when for comfort's sake the rain should be falling fallen but hasn't yet, I watch a Tuamotuan family fishing for supper. A pair of *poi* dogs comes with them, trotting confidently down the beach, stopping only to wheel on their butts and counterattack the fleas. As the father wades into waist-deep water, holding a shimmery gill net at shoulder height, the dogs watch patiently. Accompanying him is a teenage boy towing a plastic tote, the aquatic equivalent of a red wagon. Three small children follow, one boy with mask, snorkel, and fins, the other boy on a boogie board. Only the little girl is free of gear.

Wading slowly, the father examines the water. As with the *poi* dogs, he must discern the good fish from the bad, and when the signs are right, he swings the net away from him in a gesture as graceful as a tai chi pose, the net flaring out and floating before settling onto the water. As the weights drag the net underwater, the boy in mask and fins thrashes toward it, exaggeratedly slapping his arms and fins on the surface and driving the fish before him into the net. The little girl does the same but without the fins. The boy on the boogie board sweeps in from the side. The dogs sit on the beach and watch, attentive but dry. With the father anchoring one end of the net, the teenage boy walks around the thrashing children to close the circle, and, a few moments after the trap is set, the net is hauled in.

Aquiver, the dogs watch as the pretty little jewels of reef fish are untangled from the snare and tossed into the floating plastic tote, where their slapping convulsions can be heard from shore. Enchanted by the sound, one of the *poi* dogs rises from the sand to stand in the shallow water, belly deep, pointing in the direction of the hunt.

It's the evening hour. The sun is spreading its silvery blush across the water and flocks of black noddies are dancing above the two catamarans moored offshore. All the hotel guests are floating in the infinity pool. The staff are busy tending bar or setting up the restaurant. The father eyes me warily as he gathers the gill net into a shimmery ball, and I cinch my hat down so I can watch him more discreetly. I suspect what he's after: the relatively virgin fishing grounds inside the hotel's portion of the lagoon. Bit by bit he sidles across the invisible boundary, the children following quietly. I don't begrudge him the infringement. Instead, I imagine the rewards of a dinner shared among those who worked to catch it together. In my world, families never dine together anymore, let alone catch their dinner together.

With no one to notice, the father works his way across the coral heads marking the Kia Ora's portion of the lagoon. The dogs follow from shore, stopping to scentmark a hotel guest's wetsuit left drying against a palm tree. Methodically the father leads his little crew to the lucrative piers under the overwater bungalows. The fish population here is artificially high and the individual fish are artificially juicy — fattened as they are by the guests who, by way of entertainment, drop food through the little trapdoors in the bungalows' floors. Without knowing it, these guests are closing the loop on a self-sustaining system as circular and organic as a Chinese fish pond: the Tuamotuans catch fish for the hotel guests, who feed their leftovers to the lagoon fish, which in turn feed the Tuamotuans and their dogs.

The family is targeted and efficient in working its portion of the loop, and as the minutes tick by with no one (i.e., me) objecting to their pilferage, the members lighten up until the children are splashing and whooping and hollering with joy, the secret spice to all their meals, I imagine. Food is joy here. It's not fast, but intimate and familiar and part of your neighborhood.

This is further proven to me the following afternoon, out on the Zodiac, when Manu reaches his threshold of heat tolerance and pulls the ripcord on the outboard. He drives us across the pass to the village of Tiputa, where he lives. Weaving expertly among the coral heads, he ties up to a pair of mooring buoys, then shows me how to exit the Zodiac with the minimum of effort, using dead coral heads as stepping stones en route to shore. On the beach, he moves surprisingly quickly through white coral rubble as rounded and unstable as tennis balls, headed for his property on the edge of the pass, and a broad, big-leafed tree he calls *raisin de mer* in French, and *vin'e* in Tahitian. It's loaded with grapelike clusters of fruit, mostly green, but a few turning a deep maroon.

All the low-lying fruit have been stripped off by the *troupeaux* (flocks) of children, as he calls them, who swoop down on this tree when school lets out at eleven o'clock every morning. They love this tree, he says, and for the month or so that it's in fruit, his yard is crowded with noisy, happy, juice-streaked youngsters. He tells me this as we munch together, enjoying the shade of the foliage, the tart sweetness of the fruit, the breezes ambling through the pass. I am touched by his desire to have me experience this small part of his world, and by the effort he makes to reach the ripe fruit up high so that I can taste them at their best. I am also touched by the notion that there is still a place on this planet where children race from school to the pleasures of a fruit tree.

Manu had a fish-eating dog some years ago, and in a measure the size of the big heart in this big man, the dog forswore the pack and pledged his allegiance to Manu. Each morning, when he waded out to the Zodiac to drive across the pass to work, Manu left his dog tied up in his yard, in the shade of the *raisin de mer* tree. Some mornings the dog engineered his freedom and launched himself into the water in pursuit of Manu, gamely dog-paddling his way into whatever conditions the pass was delivering up that day, including the full maelstrom of ten- or fifteen-foot-high standing waves and a four-knot current ripping out to sea.

Various islanders rescued the dog on many occasions, hauling him out of the water and driving him to one shore or another. But there was no keeping this dog from Manu, until the day came when no boat happened by to rescue him. Presumably swept out to sea on the *mascaret*, he most likely became food for the silvertip sharks, and the tiny golden trevallies riding their bows.

Inshallah, says Manu sadly.

15

The Consorting Together
of Dissimilar Organisms

ALTHOUGH WE CALL IT domestication, the relationship between dogs and people is a symbiosis, which my old *Funk & Wagnalls* dictionary (1963) defines as "the consorting together, usually in mutually advantageous partnership, of dissimilar organisms." One of the cornerstones of biology, symbioses occur among more than half of all known animal species, which enjoy or withstand a symbiotic relationship with at least one other community species — a number that will doubtless rise as we learn more about relationships in the wild.

The underwater world abounds with symbiotic relationships, and the life of the reef seems unusually creative in its application of the principles of partnering. In fact, the coral edge is so rich with intertwined species that the partners cannot be divorced from each other without losing their own identities.

First and foremost among these are the corals themselves. Long considered the simplest of invertebrates, corals are composed of little more than a hollow tube, the gastric cavity, surrounded by a fringe of stinging tentacles used to capture prey. Related polyps are connected to one another by living tissue through which they share digested food, and the resulting colonies can grow to great size and age.

This deceptively simple *bauplan* (body plan) once led experts to deduce a correspondingly simple lifestyle. But today corals are recognized as far more complex life forms, having joined the elite ranks of the plant/animal hybrids. The plant part comes into the equation because most reef-building corals are inhabited by single-celled plants, the microscopic dinoflagellates (Greek *dinos:* whirling; Latin *flagrum:* whip) known as zooxanthellae. More than six million zooxanthellae inhabit each square inch of some corals, and in return for this safe home and a ration of carbon dioxide and nutrients, the algae contribute the by-products of their photosynthesis to the coral animal: oxygen for respiration, carbohydrates for food, and the alkaline pH necessary to secrete an aragonite skeleton, the backbone of the reef itself.

So profound is this trading of needed goods that one study of the stony coral *Stylophora pistillata* demonstrated that more than 95 percent of the sugar photosynthetically produced (or overproduced) by the plant part was transferred to the animal part, with the plant utilizing less than 5 percent for itself. As a result, the combined plant/animal species becomes a functional autotroph (Greek *autos:* self; *trophe:* food), able to feed itself even in the absence of prey.

The closed system of polyp and plant is defined as a mutualistic symbiosis, with both organisms benefiting, and this turns out to be so enormously productive that it becomes a primary reason why so much life is able to flourish in the oligotrophic (Greek *oligo:* scanty; *trophe:* food) waters of the tropical and subtropical oceans. Although these warm seas seem rich, they are not. Their brilliantly clear waters, while beautiful, are evidence of an impoverished condition. In contrast, the waters in the higher latitudes are clouded green by the density of phytoplankton (the primary producers), making them ecologically productive, as measured by the assimilation of inorganic nutrients, usually sunlight, into the biomass.

On coral reefs, productivity takes place largely in private. Since the coral's zooxanthellae are endosymbionts — living completely enclosed within the coral animal, rather than as free-floating plankters — the cycling of materials takes place within the body of the reef, so to speak, as if we grew corn or potatoes directly in our bellies. This method of productivity is fast and furious, and it is utilized by other marine invertebrates that partner with zooxanthellae, including some jellyfish, the giant clams, and nudibranchs (sea slugs).

A better definition of the relationship between the plant and the animal parts of the coral might be one of synergism (Greek *synergos:* working together). In its medical definition, synergism is described as the joint action of different substances to produce an effect greater than the sum of the effects of all the substances acting separately. For example, if we were to measure the worth of the coral polyp working alone at two units, and the worth of the zooxanthellae working alone at two units, we would find the worth of the pair working together at a mysterious five units.

From its theological definition, synergism is the doctrine of human effort combined with divine grace in the salvation of the soul. Perhaps synergism becomes imbued with something other than the purely physical in order for the two of the polyp and the two of the zooxanthellae to equal the five of the coral. The Taoist sage Chuangtse, writing twenty-three hundred years ago, believed that something to be the Tao, which

> unifies the parts . . . [because] the disadvantage of regarding things in their separate parts is that when one begins to cut up and analyze, one tries to be exhaustive . . . One goes on deeper and deeper, forgetting to return, and one sees a ghost . . . For a thing which retains its substance but has lost the magic touch of life is but a ghost.

According to Chuangtse's thinking, to see the zooxanthellae without the polyp or vice versa is to see only a phantom of the powerful alliance that engineers, designs, and builds the reef. In other words, the reef without its lifeforce, the mysterious missing quantity of one.

The extraordinary marriage between the animal and the plant parts of corals has recently been revised to include a third partner: a bacterium in the connubial bed. Michael Lesser from the

University of New Hampshire and a team of researchers discovered the presence of a cyanobacteria (the photosynthesizing bacteria previously known as blue-green algae) living within star corals of the genus *Montastrea* in the Bahamas. While many corals are known to glow green when exposed to artificial light at night, this glow is the product of fluorescent proteins produced by the corals themselves. But what Lesser found in the Bahamas were corals of the species *Montastraea cavernosa* fluorescing bright orange in the daylight as a result of their symbiotic cyanobacteria.

Their research indicates that, in return for a safe home, the resident cyanobacteria release an enzyme that enables the coral to convert the nitrogen in seawater into a useable form. The cyanobacteria may also provide nitrogen to the coral's zooxanthellae, which reciprocate by supplying carbon to the cyanobacteria. What at first appeared as a two-way mutualism is apparently a three-way mutualism — the ocean's own ménage à trois, with all its inherent complexities.

This discovery is the long follow-up to a pioneering study from more than thirty years ago, in which researchers first discovered two species of luminous bacteria, *Vibrio fischeri* and *V. harveyi*, inhabiting the flashlightlike organs of some fish and squid. Later research showed that bacteria inside the flashlights luminesce only upon reaching a specific population density — a social phenomenon that has since come to be known as quorum sensing. When congregating, or swarming, bacteria share information with their neighbors, communicating through hormonelike molecules in a process so complex that some researchers now argue it bears the hallmarks of language. Upon reaching a required threshold, these molecules then trigger a mass behavioral response among the bacteria and, in the case of *Vibrio fischeri* and *V. harveyi*, prompt them to fluoresce. (In other bacteria, quorum sensing leads to the release of poisonous substances, which may then trigger the onset of a previously latent disease in the host.)

Clearly, much is going on in this miniscule world beyond our daily view. Not only are bacteria exhibiting complex social behaviors and perhaps some equivalent of language, but they are doing so in concert with their symbiont zooxanthellae and their symbiont polyp — the whole of the triumvirate adding up to a veritable microcosm in a plant-animal-bacterium creature the size of an ant. The $2 + 2 = 5$ of this synergistic equation might better be written as $1 + 1 + 1 = 5$, and no doubt the weird math will not end there, as the reef is full of alliances we are only beginning to decipher.

16

It Furthers One to Cross the Great Water

DEEP IN THE INTERIOR of Rangiroa's lagoon, far from the sight of any *motu* except the miniature islet of Motu Paio, we are lethargically studying the ways of drifters. The day is hot as Hades, without a breath of wind. Every last watt of sunlight blasts off the mirrored surface of the lagoon. For hours we have been comatose with this heat, draped across our Zodiac's furnace-hot pontoons, dribbling tuna oil overboard, watching it ooze away on the currents turbid with sand. We are hoping this oil, the ambrosia of the sea, will reach the noses of the large tiger sharks hanging around in this season, snacking on sea turtles headed to Motu Paio to nest. So far we have seen neither sharks nor turtles, although a loose smack of crown jellyfish (*Cephea cephea*) is drifting by.

In keeping with our inertia, the jellyfish are barely moving, appearing through the glass of the surface like elaborate ladies' hats, complete with feathers and beads and pleats of pink (zooxanthellae-stained) satin. They are sashaying on the current, swinging backward a half-step before advancing a full step, the pinafore of their tentacles lagging coyly behind. A complete dance step — forward, backward, pinafore-catchup — takes a lazy minute or so to complete, providing us a hypnotic respite from the heat as they parade, millimeter by millimeter, toward no other destination than afar.

For lack of anything better to do, some of us slip over the side, exchanging the heat of the sun for the hot bath of the inner lagoon. The topmost twelve inches of water registers well over ninety degrees Fahrenheit, and the jellies are noticeably avoiding this heatstruck realm. We sink, too, and from below, through the grainy soup of the lagoon, the jellyfishes' finer details — the feather-and-beadlike gastrozooids (digestive polyps) and the jeweled crowns — disappear. Through the murk, they look like nothing so much as hunks of rindless watermelons, complete with white seedlings, the oral mouths.

Sightless, devoid of steering, not as transparent as they could be, and large and juicy-looking, these jellyfish have drifted upon the shores of Motu Paio at a bad time, with the numbers of jellyfish-eating sea turtles running at an annual high, as the females gather to nest, and the males wait to breed with them. In fact, while we cool off and relax and hope that no tiger sharks are following the tuna-oil scent trail just at present, we find a hawksbill turtle enjoying the ministrations of a pair of shrimp at their cleaning station on a patchy sponge-and-algae bottom. Alert, head up, flippers outstretched, she waits patiently while the shrimp tiptoe into the secret folds of her armpits. Perhaps she is pampering herself after a night of egg-laying. Perhaps she is preparing for the coming night's labor. Perhaps she is digesting a feast of jellyfish.

The jellies drift on, seemingly oblivious to the presence of a predator. One jelly glides close to the bottom, beguilingly turned inside-out, petticoats flipped open, long tentacles exposed, heading for a collision with the sea turtle at the cleaning station. But the hawksbill seems entranced by her spa treatment, and the jellyfish wafts blithely on.

Drama, however, is never far away in the sea, and within seconds a shark sidewinds into view — not the tiger shark we have been hoping to see, but a small five-foot blackfin shark (*Carcharhinus limbatus*). This species lives in murky water, nose to the ground, running erratic search patterns on imperceptible scent signals. Our visiting blackfin is following the scent of tuna oil. And because we are bored and hot and have barely shot a frame of film, we drag a large tuna head overboard and film the nearly instantaneous appearance of five more blackfin sharks, and the manner in which they proceed to play water polo with it. It's fun in a heatstruck kind of way, although the action scares off the hawksbill turtle, leaving the jellyfish free to parade unmolested.

Yves, one of the underwater cameramen, is beginning to doubt the success of a tiger shark shoot today. He is a small Frenchman whose sixteen-millimeter camera in its underwater housing seems as big as he is. Along with Yann, he is one of the French divers extraordinaire of Rangiroa, as well as the founder of the first dive operation here. Most of what any of us know about this atoll is gleaned from him, an accomplishment made more impressive by the fact that Yves, from the Alps, is as pale as a cavefish, with glacier-blue eyes and colorless hair. In fact, he possesses a kind of jellyfishlike transparency, which seems completely out of place in this equatorial, ozone-hole, reflective-hell of a topside world. Yet he has managed to survive, head swaddled in T-shirts, nose slathered with zinc oxide, sustained by long hours of underwater rewards.

The tiger sharks are here, he reassures us, pursing his sun-ravaged lips. They are always here at this season. But they may be lazy after eating turtles, he says. They may be sleeping and dreaming of tuna after smelling our tuna oil. But they will come. Today. Or tomorrow.

And so we wait, as so much of this business requires, maintaining hope without succumbing to the perils of expectation.

The fifth hexagram in the *I Ching,* the three-thousand-year-old Chinese *Book of Changes,* is called *Hsu / Waiting (Nourishment),* and is composed of the trigram *Ch'ien* (The Creative, Heaven) supporting the trigram *K'an* (The Abysmal, Water).

The idea of heaven below, water above, is generally the reverse of Western thought, although in good company with the Polynesian view. The Chinese meaning of *Hsu* is that the water in the upper trigram represents clouds not yet ready to shed their rain, a prophecy implying that we face danger and must wait for strength, metaphorically depicted as rain. We must also maintain faith in the certainty of the outcome. The strong person will persevere cheerfully in this task of waiting, while the weak will grow agitated and push against the natural order of things, forcing a false resolution.

Some on the film crew would like to do just that, motor off in the Zodiac (it would be cooler), and cast our tuna oil elsewhere. Some would like to abandon the tiger sharks and film something else altogether, something certain, like the gray reef sharks in Tiputa Pass. Yves is clearly in favor of waiting, despite the heat, the glare, the stultifying inactivity. He is resolute beneath his improvised headdress, though he grows pinker by the hour, the zinc oxide notwithstanding. All good cameramen have this gift of stubbornness. So we wait, observing the demure jellyfish parading past, likewise waiting for some prey or predator to come along and rearrange their journey.

I have an old book stowed in a plastic float bag for occasions such as this. Purchased from a bookstore in Papeʻete for an exorbitant sum, the book is a mishmash of stories, poems, and legends from the South Seas, self-published and bound in a pretty *tapa* cloth depicting a swimming *honu* (turtle). Among the unaccredited tales is one I recognize, a story of jellyfish recounted from Aotearoa to Hawaii and all of Polynesia in between. It tells of the demigod Maui, and how he was born prematurely, so unformed and grotesque that his mother wrapped him in her hair (or, depending on where the tale is told, her daughter's hair, or seaweed) and tossed him into the sea.

In some versions of this story Maui is alive when tossed; in others he's already dead. But in all cases he is quickly rescued by jellyfish, who empathize with his shapeless, embryonic self, and tenderly wrap him in their arms, tuck him into their bells, and mother him all the way to the sea god's (or goddess's or both) house, where he is hung in the rafters to revive in the heat from the divine fires. Perhaps only seagoing Polynesians, professional ocean drifters themselves, could conceive of jellyfish as hopeful creatures with mothering instincts.

It furthers one to abide in what endures, says the *I Ching.* Yet this day, waiting for tiger sharks in Rangiroa's vast lagoon, we have not abided and have succumbed instead to speeding across the water, chasing new hopes. We are on our way to the far western edge of the atoll, to a ring of *motu* forming a miniature lagoon within the lagoon, known as the Lagon Bleu.

As the fifth hexagram of *Hsu* also says, *It furthers one to cross the great water,* and on the horizon, the Lagon Bleu appears like a sapphire mirage, its luminous colors amplified through its miniature prism. Sugar-white sand *motu* are adorned with aqua-green coconut palms, which are in turn decorated with white fairy terns, whose harsh *grrich-grrich-grrich* scratch through the thunder of crashing surf on the seaward side of the *motu.* Pairs of these terns soar in courtship flight over the small lagoon, one atop the other, holding perfect synchrony an inch apart as they tack from one wing to the other.

Although we have come here to work, something about this place takes the work out of you, and for a while we wander on private rambles. The oceanside *motu* are cobbled with coral rubble, some of it well on its way to becoming beach cement with the help of the guano of the fairy terns and the Pacific reef herons guarding the *hoa.* The view from here is of a big ocean heaving to a horizon lumpy with swells. Its pull is magnetic, and, one by one, we wander to the edge to stare at its outsized force.

We have come to the Lagon Bleu for the baby sharks, primarily blacktip reef sharks (*Carcharhinus melanopterus*) — Rangiroa's prettiest inhabitants, diminutive and delicate, at less than six feet long. They are decorated with silver and gray chevrons on their flanks, and with dorsal fins highlighted in bold black and underlined in white. Like most Carcharhinidae, blacktip reef sharks are viviparous (Latin *viviparus*, bearing live young). The females journey to traditional pupping grounds to give birth in waters sheltered and shallow enough to keep their primary predators — other sharks — away.

On this hot afternoon, under a sky laden with cumulus clouds, the tiny lagoon is animated with baby sharks. At about fifteen inches long, they are perfect replicas of the adults, only softer and more rubbery-looking, and they are energetically demonstrating the attractions of bonelessness (class Chondrichthyes; Greek *chondros*, cartilage), their bodies bowing into circles with each flick of their tails.

These pups are adorable, and even the more macho members of the crew are doing a male version of cooing. Although we have come here to film these babies, all we really want to do is cuddle them. And so we do, wading in ankle-deep water so hot it makes us hop, and chasing down the little black-and-white sails of dorsal fins. Grabbing a shark behind the pectoral fins, we hoist it into the air and assess from the claspers that he's a male. We try to sweet-talk him into abandoning, for the moment, his circular swimming motion, more like thrashing, really, with a strength simply unbelievable for his size. After about twenty seconds, he surrenders, going as stiff as a board. We pet him carefully, in one direction only, since his skin is composed of dermal denticles every bit as sharp as little teeth.

But the pup does not enjoy this forced removal from his universe, and no doubt this was not one of the perils calculated into the evolutionary equation when mother sharks elected to give birth and then abandon their pups to raise themselves. So we don't pet him for long and, thankfully, he resists the urge, or maybe the right, to punish anyone's ankles when we let him go.

17

The Lemon
Shark Affair

FROM A DIVER'S POINT of view, sharks are the defining players at Rangiroa Atoll, appearing in such numbers and such diversity as to render even the most fearful adventurer blasé after a few dives. On the outer edge of Tiputa Pass, the gray reef sharks (*Carcharhinus amblyrhynchos*) spend the daytime hours schooling. On this shoot we are in search of the hordes rumored to live very deep. At about one hundred feet underwater, we find shoals of ten or twenty gray reef sharks drifting in close formation. At one hundred fifty feet, the shoals coalesce into squadrons. Below two hundred feet, they become curtains of sharks drifting slowly in the bottleneck of the deep pass.

Breathing compressed air at these depths, we are affected by nitrogen narcosis, known as rapture of the deep or the martini effect. Being narced, as divers call it, feels euphoric, something akin to the effects of inhaling nitrous oxide. Yet tempering the fun is our knowledge that all divers face a confusion threshold: a depth at which — in the jargon of diving — we will giddily hand our regulators to passing fish.

Descending past one hundred feet, we grow heavier by the moment, and jollier by the moment, a classically hazardous paradox because just as we are feeling invincible, we are, in fact, in grave danger. At depths where nitrogen narcosis develops, our compressed air is further compressed, meaning we breathe it more quickly — raising the danger of running out, even as we accelerate toward the point of no return in terms of decompression diving.

We are, however, in the words of pioneer diver Frederic Dumas, as merry as bubbles, and at one hundred sixty feet we begin to glimpse the humbling spectacle of the armies of gray reef sharks. Narcked or not, we have entered a primeval realm populated with legions of ancients, as if swimming with dinosaurs, or discovering Shiva, whose many heads, arms, and legs manifest as sinuously dangerous fish.

Much power is inherent in this sight. And yet it is quiet, or nearly so, and almost still, the sharks drifting on an invisible conveyor belt, surrounded by trusting clouds of bluelined snapper, the mobilization echoing the birth of a typhoon or a war, as all the elements of annihilation collect and await the tumult.

There is no terrestrial equivalent. Apex predators don't mass on land. Imagine herds of hundreds of tigers or packs of thousands of wolves. One reason for this arises from a fundamental difference between the two energy systems. In the topside world, herbivores (the primary consumers) eat only a meager 1.5 to 2.5 percent of the green plants (the primary producers), because trees are protected by their bark and their size and are difficult to eat (they yield eventually to the detritivores). In the oceans, however, the primary consumers, which are mostly zooplankton, eat between 60 and 99 percent of the primary producers, the phytoplankton.

Corals, with their zooxanthellae plant partners, play the role of both primary producers and primary consumers and thus contribute to the lopsided ratio in the seas. Yet even in temperate waters beyond the Darwin Point, the primary consumers exist on a far larger scale than on land, a differential that ripples through all the ensuing trophic levels, making the populations of secondary, tertiary, and quaternary consumers proportionally larger as well. Furthermore, whereas the apex predators of the land (such as ourselves) consume animal protein largely from herbivores (cows, sheep), the apex predators of the sea consume most of their animal protein from the third or fourth trophic level. They eat fish who eat fish, who eat fish, who eat fish, who eat plants. This is why more carnivores are in the sea than on the land.

Overall, there is also more room for life in the world ocean, which contains enormous floating communities. The suspended world is full of microscopic life and dissolved organic matter, and it feeds most sea life even as it provides them living space. Imagine if the air we breathe and live in fed us too. Consequently, of the thirty-three phyla of animal life occurring on our planet, only twelve occur on land, where only one is endemic. Thirty-two occur in the oceans, where fourteen are endemic. And whereas 90 percent of terrestrial life forms are arthropods (mostly insects), nine-tenths of the ocean's life falls within eight phyla. Truly water is alive.

As for why so many sharks gather in Tiputa Pass, it is, apparently, a good place to sleep. Just as it behooves us to stay at home in bed when asleep, it befits the gray reef sharks to school in their ocean bed, tucked between the sheets of land on two sides and protected by a doorway through which any potential predator must enter — a doorway the gray reef sharks lock with the deadbolt of their collective bodies.

Yet there is probably more to this gathering. Perhaps the sharks are conducting elaborate social lives unrecognizable as yet to us. Perhaps they are congregating to facilitate the kind of cooperative pack-hunting only recently observed in whitetip reef sharks. Whatever it is, this schooling is an outgrowth of the taxonomic maturity of sharks, who have been on this planet for four hundred fifty million years — three times as long as the dinosaurs, one hundred times longer than us, and long before any creature with a backbone came ashore, or any insect took to the air. In the end, the lives of sharks may simply be indescribable in our newfangled language.

Many years ago at the Hôtel Kia Ora there was a predawn wake-up call, as soft and undemanding as a caress. Half asleep, you could easily mistake it for the sound of windswept rain on your pandanus-thatched roof. But this sound had a rhythm that didn't match the wind, and that difference would eventually tug you from sleep no matter how firmly you were lodged there. The cause was an ancient Rangiroan named Toanui, toothless and smiling, wearing a battered baseball cap from New Zealand, with a grinning kiwi bird in the classic American *Keep on trucking* pose. Every morning Toanui swept the unpaved walkways with a largely debristled broom, pushing aside the night's litter of pink frangipani petals, iridescent insect carcasses, and cigarette butts. He took the coconuts that had fallen in the night and stacked them in little pyramids, offering to open them with his machete for any guests on their way to breakfast. Although he spoke only Tuamotuan, with a smattering of French, this didn't stop him from conversing at length with anyone who had the time to listen.

When I asked if he believed in Taputapua, the Polynesian shark deity, he told me a long story, complete with a more-or-less decipherable sign language, about his grandfather who, after death, manifested as Taputapua. Taking the form of a very large lemon shark, he swam all the way from Rangiroa to the neighboring atoll of Manihi to tell his son-in-law to be nicer to his wife. When I asked Toanui how he knew this, he said that he'd seen with his own eyes the large lemon shark swim through Tiputa Pass toward Manihi, and he'd heard how the same lemon shark appeared in Manihi's lagoon, menacing the son-in-law's *va'a* by bumping his *ama* (float).

Across French Polynesia many people believe that an ancestor can take the form of Taputapua and travel between islands or *motu* in times of family emergencies or arguments. Some Tuamotuans today claim to have seen Taputapua carrying a relative on his back from island to island in order to visit distant family. Because Taputapua is called upon to avenge wrongs, many spearfishermen, who are particularly vulnerable to sharks when carrying bloody fish underwater, will not work if they are fighting with their wives. And so down the centuries, the belief in Taputapua has served the sharks (and the wives) of the Tuamotus well, with many fishermen refraining from hunting or harming sharks in any way.

But this is changing. In 2003 posters appeared in the villages of Rangiroa Atoll reading NOUS ACHETONS DES AILERONS DE REQUINS (we buy shark fins) — fins destined for restaurants throughout Asia as the essential ingredient in sharkfin soup. Once a rarified foodstuff of the elite, today sharkfin soup is an affordable luxury for the Chinese nouveau riche who wish to prove their wealth by ordering a two-hundred-fifty-dollar bowl of glutinous cartilage flavored with chicken broth. At expensive eateries in Hong Kong, Taipei, Singapore, and Bangkok, middle-class diners slurp this pricey food even as the World Conservation Union adds more shark species to its Red List of Threatened Species.

Sadly, the four thousand tons of shark fins exported to Asia yearly, representing four million sharks, is exceeding the carrying capacity of shark populations. In the Gulf of Mexico, the number of oceanic whitetip sharks has plummeted 99 percent since the 1950s, driving this once common pelagic species into effective extinction. A study of the North Atlantic Ocean found that overall shark populations have plunged more than 50 percent since 1986.

Unlike the surgeonfish in Tiputa Pass, who spawn millions of eggs a night, sharks are slow breeders, with most delivering small litters (some only twins) after reaching a late sexual maturity (some at twenty-five years old), after which they typically deliver litters at three-year intervals. The results of such slow reproduction make recovery from overfishing notoriously difficult. When Europeans overfished porbeagle sharks in the 1960s, the species struggled toward recovery for the next thirty years, finally achieving some semblance of health in the 1990s, only to become the target of U.S. and Canadian fleets that fished it into commercial extinction in three short years.

En route to Avatoru Village by the one and only road on Rangiroa, I pass the dock near the airstrip and notice a fishing boat and a few *pêcheurs* hanging around talking with Yves and his wife, Sandra. By the time I return everyone is gone. Only the next morning does Sandra tell me what transpired — how they encountered the *pêcheurs* unloading a pair of large female lemon sharks (*Negaprion acutidens*) from their skiff, their fins destined to the merchant who had posted the signs reading NOUS ACHETONS DES AILERONS DE REQUINS.

Concerned, Yves and Sandra tried talking to the fishers about the realities of fishing sharks — the slow breeding, the small litters. But many Tuamotuans are sensitive about colonialists telling them what to do, and here in Rangiroa the fishers are worried about a proposal to outlaw shark fishing. Sandra, a petite, pretty blonde Colombiana of French descent, who is the executive director of a foundation overseeing a national marine preserve in Colombia, confronted one of the fishers. They quarreled. He pulled a knife, pressed it to her throat, and threatened her life.

In the coming days, *La Dépêche* is all over the story, dubbing the incident *L'Affaire des Requins Citrons* (the Lemon Shark Affair). The newspaper accounts always refer to Sandra as a *touriste,* even though she is French and married to a resident of Rangiroa. Yves is angry about the treatment of his wife and is pressing charges against the fisherman. He is equally distraught by the thought of a shark fishery taking hold in Rangiroa, telling me bitterly that unless shark-finning is outlawed he will give up the dive business and open a pizzeria.

Do you know how much a pizzeria makes in a month? he asks. Ten thousand to thirty thousand dollars, he says, steady money, not like the dive or the film business.

He looks desperate, the skin peeling from his nose and lips, the almost-certain price of skin cancer he will pay for his years of building a business from nothing here. The thought of Yves — extraordinary diver and cameraman — giving up the underwater world for a hot oven on a hot atoll seems about as cruel as the thought of a Tuamotuan *pêcheur* — born to the sea — out of business and forced to work for Yves in that pizzeria. Yet all will suffer if the ecological equilibrium of the atoll is destroyed. A study on Bahamian reefs by Mark Hixon of Oregon State University found that the removal of predators (and, interestingly, competitors) destabilizes the remaining species, eventually reducing biodiversity.

Bikini Atoll, in the Marshall Islands, attests to this. Quarantined for decades following U.S. nuclear bomb tests, Bikini was recently opened to dive tours, where its primary draw was an enormous population of resident gray reef sharks. Unaccustomed to and unafraid of us, these sharks proved somewhat more aggressive than their Tuamotuan counterparts. But sports divers love aggressive sharks, and Bikini quickly began to generate foreign revenue for the Marshallese. That is until illegal long line fleet from Taiwan and Hong Kong stripped the atoll of all but a handful of surviving sharks. Today, Bikini offers pretty reefs and a titillating radioactive factor (a bragging point for the ultramacho set), but few sharks, less biodiversity, and increasingly fewer divers.

Without its sharks, Rangiroa will likewise be perceived as a site with an exciting drift dive and not much more, and hardly worth the brutal airline hours and bruising dollars it takes to get there.

18

Grand Secret

T HE TUAMOTUS OFFER their own version of Bikini. Moruroa Atoll (Mangarevan *moru:* secret; *roa:* grand) and neighboring Fangataufa Atoll are the last islands in the Tuamotus chain. They lie far enough from Rangiroa (six hundred fifty miles to the southeast) that they seem more a part of the easternmost of French Polynesia's archipelagos — the Mangareva Islands, which the French call the Gambiers. Because of their extreme remoteness, Moruroa and Fangataufa atolls were chosen as the site of the more than two hundred (the number is contested) nuclear detonations the French conducted between 1963 and 1996 at their coral-reef test range, the Centre d'Expérimentations du Pacifique.

No one who wishes to avoid arrest and prosecution under French military law visits either of these atolls today. No sailing charts are available, and the few maps and photos around show Moruroa as a lovely atoll, its sand *motu* assuming the shape of a snail, with two passes at the base of the snail's shell, only one of which, apparently, is navigable. Squinting at the map, you might imagine the high island that once stood there, seventeen miles wide by seven miles long. Not a large island, by any means, but one formerly graced with at least two rivers. What remains, as seen in a few aerial photographs, are sand *motu* laden with prefabricated military warehouses and acres of airstrips.

Of the one hundred eighty-one acknowledged nuclear tests conducted in Moruroa and Fangataufa (including a few two-hundred-kiloton bombs ten times more powerful than the Hiroshima bomb), forty-four were exploded in the air. These atmospheric tests continued for a full eleven years after the United States, Britain, and the USSR signed the Partial Test Ban Treaty in 1963, banning atmospheric nuclear tests.

In fact, France detonated its first bomb at Moruroa three years after the PTBT. The bomb was exploded aboard a barge in the lagoon, sucking water into the air and raining dead fish, corals, cephalopods, crustaceans, mollusks, and all the once living components of the reef onto Moruroa's *motu,* where their radioactive forms decayed for weeks. Confounded by this result, the French hastily arranged to explode their second bomb seventeen days later from an airplane forty-five thousand feet above the featureless South Pacific, some sixty miles south of Moruroa. Without people or equipment to witness, record, or analyze this distant blast, virtually no data was collected, making its detonation more an act of pique than science. Two days later, as described by the *Bulletin of the Atomic Scientists:*

> an untriggered bomb on the ground [at Moruroa] was exposed to a "security test." While it did not explode, the bomb's case cracked and its plutonium contents spilled over the reef. The contaminated area was "sealed" by covering it with a layer of asphalt.

The French conducted the next test eight weeks later, on September 10, 1966, with President Charles de Gaulle in attendance aboard a warship equipped with iron shields and decontaminating sprinklers. The blast was postponed for a day because the wind was blowing west toward inhabited lands instead of south toward Antarctica. When these unfavorable conditions persisted on the second day, De Gaulle protested against further delay, citing his busy schedule. And so the bomb, hanging from a helium balloon, was detonated despite the wind. The fallout hit the island of Tahiti hardest, with radioactivity spreading across the South Pacific to the neighboring islands and nations of Niue, the Cook Islands, Tonga, Samoa, Fiji, and Tuvalu, as registered by the New Zealand National Radiation Laboratory.

The remaining forty-one atmospheric tests the French conducted delivered such high rates of fallout to the Mangareva Islands that the French radiological security service advised evacuating the islands of Mangareva, Pukarua, Reao, and Tureia. But this was never acted upon. And so, as with the Marshall Islands, which withstood sixty-six nuclear bombs detonated by the United States (twenty-three on Bikini and forty-three on Enewetak), the islands of Polynesia and their inhabitants began to absorb and suffer the effects of a persistently radioactive environment. The people were advised not to eat the fish or the coconuts, a suggestion few could follow because these were and are the staple foods of the region.

Perhaps the strangest and saddest effect of these tests are the birth defects referred to throughout the tropical Pacific as the jellyfish babies — infants born either full-term or premature who possess no eyes or heads or limbs, and bear so little resemblance to human beings that their mothers, or the midwives who care for them the few moments they exist before dying, refer to them as little jellyfish.

Only in 1974, after more than a decade of outlaw status, and after New Zealand and Australia began proceedings in the International Court of Justice at The Hague, did France, under president Giscard d'Estaing, proclaim an end to atmospheric testing. The pronouncement initiated a new phase of detonations carried out in underground tunnels drilled sixteen hundred to four thousand feet beneath the atolls. These were the only bombs ever exploded beneath coral atolls, and the effects were profound. A 1979 accident left a bomb stuck halfway down a thirteen-hundred-foot shaft. Unable to dislodge it, the military detonated it where it lay, triggering an island collapse. Thirty-five million cubic feet of the outer atoll slipped away, generating a tsunami that wreaked havoc on Moruroa, injuring several workers, and leaving a crack more than a mile long across the atoll. The effect of the release of radionuclides into the ocean has never been assessed.

By 1981, eighteen years after testing began, the enormous mass of Moruroa Atoll had subsided five feet, shrinking an average of seven-tenths of an inch with each detonation. Alarmed at the rapidity of the island's demise, the French switched to underground testing inside Moruroa's lagoon, hoping to be closer to the center of what they imagined to be a stable core. But that same year, for the first time since 1906, a series of cyclones swept through this part of the Pacific, producing such devastating waves that the civilian technicians employed at Moruroa leaked a secret report to the French press, revealing that forty-five pounds of plutonium sealed with asphalt — the waste of security tests gone awry — had been washed away. The storm surge peeled off the tarmac Band-Aids and exposed thirty-two thousand square feet of contaminated sand to the lagoon and to the waters of the open sea.

From there the plutonium traveled who knows where. The French never tracked it, as they never tracked any of the fallout. And whereas the United States has paid $759 million, including nonmonetary compensation, to the Marshallese since 1956 (not enough) and has continued to monitor ongoing health problems (although not vigorously enough), the French have never acknowledged any problems. Cannily, they bought secrecy years ago when they required all employees of the Centre d'Expérimentations du Pacifique to waive all rights to speak, to access their own health records, or to sue for future medical problems. Underscoring such ruthlessness, in 1985 French intelligence agents, at the behest of President François Mitterrand, sunk the protest ship *Rainbow Warrior,* en route to Moruroa, killing a Greenpeace photographer in the process.

Some of the forty-five pounds of radioactive plutonium from Moruroa washed into the sea by the cyclones may have traveled the currents that ply the Tuamotus, arriving at Rangiroa's tranquil lagoon. Sitting in the restaurant on a quiet, off-season morning, as the sun creeps up the coconut palms, shaking their shadows across the ripples on the lagoon, I can't help but think of Moruroa's afterlife.

In 1976, at the height of nuclear testing, the World Health Organization (WHO) published the only study ever conducted of radiation exposure in French Polynesia. The target of their research was ciguatera fish poisoning (CFP) — a potentially fatal condition acquired by eating fish infested with the toxin of a species of dinoflagellate named *Gambierdiscus toxicus,* after the Gambier (Mangareva) Islands.

Although ciguatera poisoning has occurred along tropical coastlines for centuries, in the late 1970s — at the time of the WHO study — it had become a serious problem in French Polynesia, affecting a growing number of victims with a frightening constellation of symptoms, including vomiting, diarrhea, headache, blurred vision, fever, trembling, numbness, tingling, itching, inversion of the senses (hot feels cold), loose teeth, pain on urination, joint pain, slow heart rate, arrhythmias, low blood pressure, paralysis, respiratory failure, and circulatory collapse. By 1976, two years after the last atmospheric tests, the number of cases in French Polynesia exceeded the number of cases in all the remaining South Pacific islands combined.

Along with acute symptoms, there are many chronic responses, including itchy skin, malaise, depression, headaches, muscular aches, and odd feelings in the extremities that can persist for weeks or months. For those suffering the long-term effects, a relapse can be triggered by eating nuts or fish — even noninfected ones — or by drinking alcohol or caffeine. Additionally, the toxin is cumulative, so repeated exposures are increasingly severe. To complicate matters, CFP is also transmissible through sexual contact, through breast milk to newborns, and through the placenta to the fetus, where it can trigger premature labor and spontaneous abortion.

Although of little concern to the temperate world, ciguatera poisoning is a serious health risk for people in the tropics, and the WHO study found it flourishing wherever reef-building corals had died and been replaced by algae — conditions rampant in eastern French Polynesia during nuclear testing. Zeroing in on the islands hardest hit by CFP epidemics, the study pinpointed those closest to Moruroa, including Mangareva, Reao, and Pukarua, which the French radiological security service had advised evacuating. This was where radioactive fallout was highest, where the French military cleaned its contaminated ships after each test, and where most of the coral reefs were dead as a result.

The French have always officially denied that the epidemics of CFP are in any way connected to the nuclear test range in Moruroa and Fangataufa. They have also studiously avoided efforts to investigate the leukemias, brain tumors, and thyroid cancers — among other cancers — afflicting French Polynesians since the 1980s. Assiduously, they have kept Polynesian cancer statistics secret, or else veiled them in a bureaucratic quagmire as effective as secrecy.

The world breathed more easily after 1991, when the USSR and the United States halted nuclear testing and the French declared a moratorium. But in 1995, the newly elected French president, Jacques Chirac, declared that France would — against the grain of history — resume testing. On September 5, 1995, the first of a new test series of eight bombs was exploded beneath Moruroa. Thousands of French Polynesians turned against their own capital city of Pape'ete in the worst riots in that nation's history, burning much of Fa'a'a International Airport and large swathes of downtown. The worldwide protests that followed, along with boycotts of French goods by Australia and New Zealand, and the ignition of an independence movement among native Polynesians, finally led Chirac to declare that France would conduct only six of the planned eight tests — a number he stubbornly abided by. Only on January 27, 1996, was the last nuclear bomb exploded in the South Pacific beneath Fangataufa Atoll.

Meanwhile, poor, shattered Moruroa, once a lovely gem in a tropical sea, is now so riddled with cracks, fissures, submarine landslides, and subsidence that it may well be leaking radioactivity into the biosphere. A computer-generated model developed by scientists from New Zealand predicts that radioactive groundwater seeping through fissures in the atoll at the rate of three hundred twenty-five feet a year will leech into the Pacific by 2020. A 1990 Greenpeace study found cesium-134, a by-product of nuclear fission, poisoning plankton outside Moruroa's twelve-mile exclusion zone, indicating that such contamination may have already begun.

When the Hindu god Krishna, the eighth incarnation of Vishnu the Preserver, was born, he escaped a rash of infanticides under way at the command of the king of Mathura by being whisked to safety across the Yamuna River, where he was raised by mortal parents. His adoptive mother, Yashoda, was largely unaware that her mischievous son was a divine being. Like the Polynesian demigod Maui, abandoned as a jellyfish baby by his mother, Krishna escaped death to live a boyhood full of pranks and adventures, and Yashoda grew accustomed to confronting him over his playful misdeeds.

After his friends came to her one day and said that Krishna had been eating mud, and Krishna subsequently denied the charge, Yashoda, like any good mother, asked her son to show her the truth by opening his mouth. There inside, for the brief moment he opened up, Yashoda caught a glimpse of the universe — the entirety of time and space, of galaxies, of the birth and death of stars, of everything that ever was or ever would be — contained between her son's teeth.

In the coral atoll is a place where the sea stops being the sea and becomes the lagoon. It doesn't matter where the coral atoll, or what its name, as long as it encompasses the remains of a fallen giant of an island. Somewhere inside this coral ring, at that indefinable margin where the sea becomes the lagoon, you might take a moment to tarry on your dive, and, like Yashoda, glimpse a snapshot of the universe. To your left is the path to the open sea, the pelagic zone, the profound (Latin *profundus:* deep) realm where darkness incubates our planet's most enduring secrets. To your right lies the shallow lagoon, as dazzling as a stage set made of mylar and lit with lasers, home to the ever-smaller, the ever more complex, the foundational building blocks of life. In its own way, this is the inside of Krishna's mouth: the launching point for all the cycles of time, and the place to which they return.

Even at Moruroa Atoll, inside that broad pass at the base of the snail's shell, we might tarry on a dive and see the inside of everything, because Moruroa is also part of everything: the radioactivity, the shattered coral, the secrets. The components of life that broadcast from its pass on moonless nights, when its surgeonfish gather to spawn, may well be contaminated components. And the miraculous moments of conception, when contaminated sperm joins contaminated egg, may well produce something that the people of the nearby atolls should not eat, although they probably will. Yet even this is part of the inside of Krishna's mouth — one of the vignettes of eternity that Yashoda surely must have glimpsed.

In the Indian state of Bengal, where half my family roots begin, Krishna is worshipped as Jagannatha, the Lord of the Universe. His home is the Temple of Jagannatha in Puri on the Bay of Bengal. Each June, his image — a squat, stumpy, oversized head with arms coming out of the ears — is taken on an extravagant tour of the countryside aboard a huge chariot. Built from scratch each year, Jagannatha's chariot measures roughly fifty feet high and forty-five feet square and sits astride sixteen wooden wheels seven feet in diameter — the mobile home of the Lord of the Universe.

The chariot is drawn by forty-two hundred professional pullers, but is helped along by hundreds of thousands of devotees, all of whom strive to cleave the crowd and catch a glimpse of

Lord Jagannatha (very auspicious) or, better yet, to touch the chariot or one of its hauling ropes (most auspicious). Surrounded by a sea of the feverishly ecstatic, the chariot sways like a wild elephant — or so thought the seventeenth-century poet-saint Salabega — with Jagannatha vaporizing in a flash all the sins of his believers, no matter how grievous.

Making its way two miles down the Grand Avenue, the throng carries Jagannatha to the Gundicha temple. Along the way, most years, a few unlucky disciples are crushed beneath the wheels of the chariot, while others throw themselves under (or once did), desiring deliverance from the cycle of birth and death and the subsequent bliss of nirvana. Upon witnessing Jagannatha's monstrous chariot, its unhurried and invariably deadly path, the British coined the word *juggernaut.*

My old *Funk & Wagnalls* defines *juggernaut* as any slow and irresistible destructive force, and this might be the best description for Moruroa Atoll and its tortured innards — once the place where the earth's molten interior rose to the surface to make new land, now the place where science creates itself. Like a fetus feeding on its own glowing yolk, it grows bigger, perhaps big enough to take over the sea around it: the bellybutton of a new kind of world.

But it's a lovely morning in Rangiroa, the sun is rising behind the Hôtel Kia Ora, casting the purple shadows of coconut palms onto the gray lagoon, with its ripples of wind or feeding mullet. A few noddies are on the wing, and a few fishing skiffs are carving deep tracks into the water, flushing schools of trevallies into the air. Two French couples are already reclining on chaise longues on the edge of the infinity pool, talking softly, sipping *café,* as the first rays of sun slip out to meet their naked skin. Both men and women are wearing only miniscule bathing-suit bottoms, and so the first sensations of the day — the sun, the breeze, the moisture from the sky — are washing over them in the equivalent of an atmospheric massage. These four are on vacation in paradise, and yet their faces are reserved and unsmiling.

At Tiputa Pass, the *mascaret* is building, beginning as a vague restlessness at the surface, as if someone is tugging at the fabric of the water from underneath. In an hour or so, it will rage with its full force into standing waves. The Zodiacs are already gearing up. The drivers are stashing scuba tanks aboard. We are stuffing our goody bags with the objects that enable us to survive in the underwater world, the fins and masks and spare regulators. I am scrubbing my dive slate with beach sand, trying to erase yesterday's cryptic notes. *Orange space,* or *speck,* it says (though already I cannot remember what that meant), along with a doodle of what might be a barracuda. Even though I scrub and scrub, fragments of the words, along with half the tail of the barracuda, remain with me.

Part II

Funafuti
Tuvalu

What blossoms
yet has no fruit
is the white wave of the reef
putting on
the sea god's head.

— *Ono No Komachi*

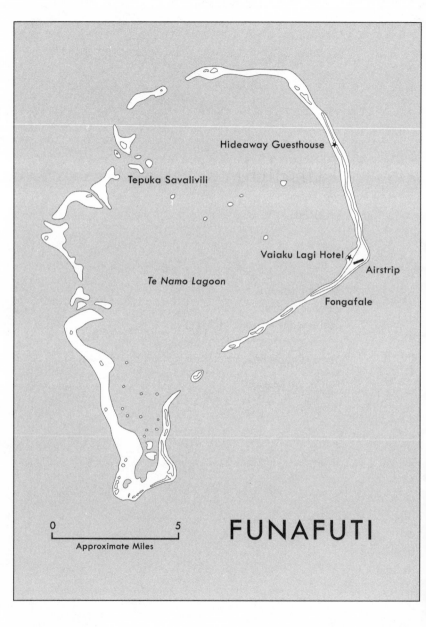

Hideaway Guesthouse ✱

Tepuka Savalivili

Te Namo Lagoon

Vaiaku Lagi Hotel ✱
Airstrip

Fongafale

0 _____ 5
Approximate Miles

FUNAFUTI

19

Hideaway

OR A FEW MOMENTS before landing, from the myopia of altitude, Funafuti Atoll appears as a lovely place, albeit a broken place — a broken pearl necklace scattered on the blue throat of the tropical sea. Unlike the Tuamotu Archipelago, where companionable, neighboring atolls are always in sight from the air, no other land is visible here, only an ocean without end, and its own billowy breath rising in cumulus clouds seemingly more substantive than the tiny land below. On final approach, this atoll assumes the classic dimensions of a desert island: a sand outpost studded with coconut palms, surrounded by wave crests longer than the island is wide.

Charles Darwin observed that "low, hollow coral islands bear no proportion to the vast ocean out of which they abruptly arise; and it seems wonderful that such weak invaders are not overwhelmed by the all-powerful and never-tiring waves of that great sea." Although he eventually discovered the reef-building mechanisms of corals that prevent atolls from succumbing to the waves, even Darwin's prescient mind could not imagine the dread possibilities of the twenty-first century — that the sea might rise faster than the corals could fortify themselves, and that these fragile spits of sand might disappear beneath the waves that tossed them into being in the first place.

Yet this is precisely the problem facing Funafuti. Like its eight sister atolls, which comprise the tiny South Pacific nation of Tuvalu, this islet rises only twelve feet above sea level. Ten thousand people live on this low ground, and their future is inextricably tied to the future of their coral reefs, and to the shifting hydrography of the Pacific. As bad as their fate may be, it represents only a fraction of a percent of the roughly one million people living on coral islands worldwide, and an even tinier percentage of the hundreds of millions living on low-lying real estate equally vulnerable to the rising waves. At risk are unique human cultures, born and bred in watery isolation. Tuvalu's government refers to the threat facing its people as a form of creeping terrorism.

Once the Ellice part of the British Gilbert and Ellice Islands, Tuvalu comprises one of the smallest and most remote countries on earth, possessing a combined landmass of ten square miles scattered over 347,400 square miles of ocean — less than half the square footage of Manhattan sprinkled across an area of ocean larger than California, Oregon, and Washington combined.

By modern measurement, Tuvalu lies between five degrees and eleven degrees south latitude, four degrees west of the International Date Line — in other words, close enough to the doldrums to keep the tourist pioneers, the tradewind sailors, away. By ancient measurement, this archipelago was one of the many island-hopping routes connecting the powerful kingdoms of Polynesia, falling midway between Hawaii and Aotearoa, roughly one thousand miles north of Tonga, and twenty-two hundred miles northwest of Tahiti.

From Tuvalu's capital, the atoll of Funafuti, as the *huao* (albatross) flies, Rangiroa lies about twenty-one hundred miles to the southeast, with plenty of rest stops available at Uvea, Nuku'alofa, Apia, Suwarrow Atoll, and Bora Bora. These island neighbors hint at Tuvalu's ethnic heritage, with 95 percent of the people descended from Samoans, Tongans, and Uveans, who arrived two thousand years ago, at a time when sea levels in the Pacific fell and the atolls of Tuvalu rose. Today's Tuvaluans share the same basic language as their Polynesian forbears, although a form noticeably gentler than Tuamotuan, with the glottal stops slurred to little more than sighs, as if the people here are perpetually longing for something.

My flight arrives from Suva, Fiji, aboard a rickety twin-engine turboprop, which an Australian aboard tells me was patched with bubblegum before takeoff, after he pointed out liquid leaking from one of the wings. The antique aircraft is peopled with Polynesians and Fijians so hefty that all the seats are fitted with seatbelt extenders. I am not overly concerned, having spent that worry on a seminal flight from Manihi Atoll in the Tuamotu Archipelago years ago, where the seatbelt extenders were not large enough to accommodate the enormous members of a touring Samoan band, each of whom required two seatbelt extenders. Adding to their weight was the fact that only half of Manihi's airstrip was operable, the other half blocked by coral clasts the size of school buses washed ashore in a recent cyclone. Yet all of us, the film crew with twenty-five cases of (overweight) film gear, plus the Samoans, survived.

Now, bubblegum notwithstanding, we touch down at Funafuti International Airport to the sound of a blaring klaxon designed to scare the pigs, dogs, chickens, and villagers off the airstrip. The ocean breeze is strong and constant on this part of the atoll, where the Americans built a runway during World War II, and many living things come here to nap in the afternoon heat, ambling off reluctantly when the little plane from Fiji — Tuvalu's only flight — comes in once a week.

Despite the fact that the airport is composed of a one-room, open-air terminal, and despite the fact that we are the only flight between now and one week from now, there is still the requisite wait for our luggage — in this case because the Tuvaluan baggage handlers are taking the time to greet their compatriots disembarking from the plane. Although most of these returnees are coming only from Fiji, seven hundred miles away, they are greeted with as much affection as if they had just completed a *vaka* (Tuvaluan: canoe) voyage to Rapanui and back. It's a charming scene, though one more restrained (thanks to the Christian Church of Tuvalu) than among Rangiroans. Here the three Gallic kisses are replaced by formal handshakes.

Observing me alone and unmet, a Tuvaluan woman, who introduces herself as Emily, asks where I'm staying. I have been watching her since our six-hour flight delay in Suva, where she was greeted with deep respect by the other Tuvaluans in the departure lounge. I tell her that I'm hoping for a room in the one and only hotel in the nation of Tuvalu, but have not been able to reach it by phone or fax or email to confirm this. She says she would be happy to have me stay with her at her home, the Hideaway Guesthouse, six miles out of town, but she's too tired just now, as she has returned all the way from London, so could I come out in a couple of days' time?

She points the way to the Vaiaku Lagi Hotel. Sidestepping the piglets on my way out the front door of the airport, I set off with a backpack and duffle bag of dive gear. It's a heavy, slow, hot slog down the sandy street and around the corner, and when I arrive at the hotel, I'm more than ready for a shower and a bed. The pretty young woman at the desk is happy to see me, asking hopefully if I'm a tourist. Not really, I say, which seems to sadden her. Her smile diminishes more when she informs me that all sixteen rooms of the hotel are full, and five people are sleeping on the floor of the conference room.

We can't have you sleeping on the beach, she says. So she calls around to the few guesthouses in town, which are also full, and then proceeds to call her aunties, a list that lasts a full eleven phone calls, speaking to each in Tuvaluan, though each call ends in disappointment, no matter the language. There's a wedding, she explains, and all the relatives have come in from the outer atolls.

She makes one more call and her smile blooms again as she

tells me the hotel driver will take me to a place down the road where I can sleep after my long trip from Australia.

America, I say.

America? She looks surprised. In these parts, bereft of television or radio or movies, America is a bit player, barely visible on the radar screen of the blue Pacific.

And so I arrive at the Hideaway Guesthouse, despite the fact that Emily is more exhausted than I am, having endured some sixty hours in airplanes and airports en route from London. She is exhausted with that weariness that only international travel from a northern hemisphere winter to a tropical ocean can produce, as all the energy locked into shivering and sleepless muscles releases in the steamy sauna of a Pacific-island monsoon. She would like nothing better than to fall into bed, no doubt. Instead she is sweeping the floor in the only guest room at the lodge, even though I tell her it's not necessary, that I'm too tired to mind sand on my feet after my own long hours of travel. But it's a universal female response, as well as the primary housekeeping chore of the Tuvaluan woman. And so I help her spread the clean Gauguin-meets-Pollock *lavalava* (sarong) across the mattress, and listen to her briefing on the quirks of the bathroom plumbing.

Not long after I've experienced these quirks firsthand, and abandoned the idea of a shower before sleep, Emily's husband, Rolf, is parting the curtain into my room, stumbling as he eases himself onto my clean bed. He speaks with a heavy German accent slurred with the smell of beer, and wears a threadbare T-shirt that says *In Dog Years, I'm Dead*. Blood oozes from his legs onto my bedclothes. Och, he says, waving his hand dismissively. It's nothing. I fell off my bike. Damn bike.

I gather from the fragments of Rolf's thoughts that the bike was "finished" after this fall and that he got back to his home in the Hideaway Guesthouse thanks to a member of Emily's family in the village, who gave him a ride the six miles out of town. Although most of the rest of what he says is incoherent, I piece together a few references to the fact that a car may have been involved in his fall from the bike, and notice that in addition to the bleeding wounds on his legs, he bears abrasions on his forehead and the top of his head. Perhaps he isn't drunk. Perhaps he is suffering from a head trauma. I contemplate waking Emily for the second time, but the smell of beer on Rolf's breath woos the tired and irresponsible part of me into hoping that he will simply go away.

Night falls not long after he leaves, and I lie on the bed, staring out the open doors across the fifteen yards of the Hideaway's front yard, across the one-lane road, through the fringe of coconut palms, to the lagoon. I can't sleep, haunted by Herman Melville's advice that it's better to sleep with a sober cannibal than a drunken Christian. My room is a sauna bereft of fan, mosquito netting, mosquito coils. Four times I've upended my bags in search of a bottle of jungle juice, and have now finally resorted to lying on the bed, feet pointed toward the lagoon, listening to a melancholy Beck CD on my Discman so I don't have to hear the hordes of mosquitoes partaking of the blood feast. By the light of my headlamp I lift my arm to read, one Japanese tanka at a time:

> Although the wind
> blows terribly here,
> the moonlight also leaks
> between the roof planks
> of this ruined house.

By five in the morning I'm awake again, confused and disoriented, with the headlamp and earphones still clamped to my head, the batteries of the CD player dead, and my entire body swaddled in the *lavalava*. A tiny orange kitten has joined me, nestled in my armpit. There is not even a hint of light and yet the roosters up and down the *motu* are possessed to crow, and I hear the first murmurs of what will come to be Tuvalu's morning soundtrack: the sibilance of brooms as all the housewives sweep away the night's debris, the squealing of piglets, the *ulilili* calls of wandering tattlers (sandpipers) plying the lagoon, and Rolf, rooting through the kitchen, muttering in German.

To my dismay, a church next door begins to ring its bells, a startlingly full-bodied sound that peals like thunder chimes, triggering a cascade of bells from churches up and down this tiny *motu*. Fumbling for new batteries, I feed the CD player, set the drone of Beck to an infinite loop, and wrap myself back into the shroud of the *lavalava*, hoping for a sleep dreamless enough to last until the crepuscular mosquitoes have feasted and gone to bed for the day. Underlying the insomnolent heat, the insects, the pealing bells, and the piglets, are my queasy nightmares of a watery inundation.

20

Falling Dominoes

D ARWIN WAS ECSTATIC about the power of the tiny coral polyp to overcome the forces of the ocean, and after visiting the atolls of the Cocos Islands south of Sumatra in the Indian Ocean, he wrote, "Let the hurricane tear up [the reef's] thousand huge fragments; yet what will that tell against the accumulated labour of myriads of architects at work night and day, month after month? Thus do we see the soft and gelatinous body of the polypus, through the agency of the vital laws, conquering the great mechanical power of the waves of an ocean which neither the art of man nor the inanimate works of nature could successfully resist." Yet he also found evidence of the demise of atolls and the livable space they represent:

> At Keeling atoll [in the Cocos] I observed on all sides of the lagoon old cocoa-nut trees undermined and falling; and in one place the foundation posts of a shed, which the inhabitants asserted had stood seven years before just above high-water mark, but now was washed daily by every tide . . . [The] inhabitants of parts of the Maldiva archipelago know the date of the first formation of some islets; in other parts, the corals are now flourishing on water-washed reefs, where holes made for graves attest the former existence of inhabited land.

The same kind of attrition can be seen on Funafuti Atoll, where the shreds of sand are likewise beginning to dissolve.

Their fate seems linked to events far away, particularly to Antarctica's Larsen Ice Shelf, which has been self-destructing since 2000, calving monstrous icebergs from an icesheet at least ten thousand years old. Most scientists agree this disintegration is a result of human-induced global warming.

In fact the polar regions have become the hotbed of global climate change, with the Southern Ocean producing ever more spectacular evidence of ice collapses, and the Arctic Ocean melting like the proverbial snowball in hell. In 2002, the glaciers covering the Arctic and Greenland shrank by a record four million square miles, an area larger than all the continental United States. In 2005, only 2.06 million square miles of sea ice formed, the lowest ever recorded, and 20 percent less than the average recorded since 1978. At the same time, average air temperatures across the Arctic rose more than five degrees Fahrenheit above the fifty-year average. Meanwhile, Alaska's thawing glaciers are now adding twice as much freshwater to the world's oceans as the entire Greenland icecap. Overall, the rate of the ice melt is accelerating sharply, and the presence of more water and less ice on the planet is changing the shape of the earth, squeezing it oblate rather than spherical.

Yet the growing tome of scientific evidence apparently does not carry enough weight to tip the scales of political will, at least not from the nations whose fossil-fuel consuming habits make and keep them globally powerful. The United States and Australia both refused to sign the 1997 United Nations Kyoto Protocol, with its call to the developed world to reduce greenhouse gas emissions. Not coincidentally, Australia is the world's highest per capita greenhouse gas emitter, followed closely by the United States, the largest overall polluter on earth.

In light of the evidence, Australia continues to trumpet the fact that sea levels in the Pacific are not rising, quoting a report from the Australian National Tidal Facility, which monitors a network of tidal gauges across the Pacific, including one on Funafuti Atoll. Yet much less noted is the evidence from the University of Hawaii's tidal gauge in Tuvalu, which has been recording sea levels for nearly three times as long as the Australians' and has found that maximum sea levels are increasing at a considerably faster rate than what the Australians report. Aware of the political forces trying to undermine the science, Tuvaluan prime minister Saufatu Sopoanga says, "Here in Tuvalu we don't need to refer to reports because we see the evidence with our own eyes everyday."

Not long after dawn, the heat rouses me from my unkind bed, and I wander to the lounge of the Hideaway Guesthouse, which is also Emily and Rolf's living room and dining room. For a moment, Rolf, with a can of Victoria Bitter beer raised to his lips, looks startled to see me, and I realize that he remembers little or nothing of last night's meeting on the edge of my bed. Still, he recovers quickly, tells me that Emily is sleeping in, and offers to cook me breakfast. What would I like? Eggs? Yes, they have eggs, but not much else. Damn it. Damn boats.

I tell him I'll be happy to cook myself if he just points the way. But he insists on fulfilling his host duties (no, you're the guest, for Christ's sake), even though he's wobbly on his wounded, spindly legs, painted lavishly this morning in iridescent purple mercurochrome. He sits me at one of the two large tables in the lounge, brings me a cup of tea (Och, no milk, everything is finished!), and a solid forty-five minutes later appears from the kitchen bearing burned onions, eggs sunny-side up, and bread. I am hungry, but the undercooked eggs nearly defeat me.

Rolf is showing me a photograph of him and Emily thirty years ago, the two of them looking young and hip, Rolf with his Peter Sellers hairdo, Emily in a mod A-line minidress, Bond-girl hair, and big-eyed charisma. The young Rolf is harder to reconcile with the current man. Even though Emily has grown stout and has shorn her hair, her strength is still apparent, whereas Rolf has deteriorated into an emaciated wreck, hiccupping perpetually, rheumy-eyed, and weak as a kitten.

Look, it's no more, says Rolf of the photograph. It's finished.

As when speaking of everything from his past, his demeanor is dismissive. But there's no evading the sadness of his lost vigor, their lost youth. He tells me how he came to the South Pacific at nineteen, after the Australian government trolled Hamburg in the dead of a January winter, screening films of sunbathers on Bondi Beach, in an effort to recruit white immigrants Down Under. Rolf's father, a World War II veteran of Stalingrad, advised him to get out and see the world. After a period wandering Australia, Rolf joined the British Phosphate Commission (BPC) and spent the next eighteen years strip-mining the guano deposits on the eight-square-mile island of Nauru, one thousand miles northwest of Funafuti. Emily arrived from Tuvalu in 1972, hired by the BPC as a nurse.

Rolf was a mechanical engineer (which he pronounces *enginyuure*), who oversaw the power station supplying the lights and electricity needed to run the mining operations three shifts a days, three hundred sixty-five days a year. He had enough clout to negotiate a raise for Emily — who was making only ninety-five cents an hour to a white nurse's four dollars an hour — by threatening to have her work as a housecleaner for a dollar fifty an hour. The BPC couldn't sack him, he says, because he was a mechanical *enginyuure,* and they were hard to come by or to keep on that hellhole of an island. The commission finally offered Emily three dollars an hour, but not long afterward began to import nurses from Hong Kong and the Philippines for the ninety-five-cent price.

Rolf is on his third beer since breakfast, hiccupping between bouts of painful belches, his purple-painted legs swelling with what looks like an infection. He tells me of the German captain who visits on a cargo ship a few times a year and brings him the foods he misses: *pinkelwurst,* canned *blutwurst,* canned head cheese, German pickles, and brandy chocolates. He speaks lovingly of this food, describing how it's so much better than fish, fish, fish, fish. Yet it's difficult to picture him eating anything at all. He says that food makes him hiccup, and all he ever eats nowadays is thin soup.

It's also difficult to think of what really became of his father's advice, and whether Rolf has actually seen the world. After landing in Nauru for eighteen years, followed by Funafuti for nineteen years, Rolf's entire adult life has passed on two miniature islets with a view of little more than the sea. Four years ago he was sent to wintertime Berlin by the Tuvalu Tourist Office to hunt for tourists, but no German could be convinced to come to faraway Tuvalu, with an airfare from Fiji to Funafuti costing nearly as much as that from Frankfurt to Fiji. Despite Funafuti's international airport code — FUN — virtually no vacationers make it to these islands. Rolf returned home defeated and frozen, vowing never again to revisit his cold beginnings.

To what / Shall I compare the world? wrote the eighth-century Buddhist priest Sami Mansei. It is like the wake / Vanishing behind a boat / that has rowed away at dawn. And so it must have seemed to the inhabitants of Funafuti on the October morning in 1972, when Cyclone Bebe swept through, snapping off every coconut palm, flooding all the buildings, wrecking boats and ships in the harbor (the remains can still be seen in the lagoon), and killing six people in its 150-mile-per-hour winds. Bebe's waves tossed coral rubble onto the windward side of the atoll, creating a rampart that still stands as the highest point on the *motu* today.

Now colonized by coconut palms, pandanus, and breadfruit trees, this humble semblance of high ground is where I like to sit in the afternoons and watch the sea. As each wave climbs and withdraws, it rolls the coral rubble back and forth. The chattering these stones make is like the noise of thousands of falling dominoes, sharply audible even above the pounding surf. The precariousness of dominoes seems an apt metaphor for Tuvalu's fate, where changes to either sea levels or the coral cover will likely result in the entire nation succumbing to what Darwin described as the "irresistible power" of the "miscalled Pacific."

Snorkeling in the lagoon (there are no compressors in Tuvalu for scuba diving) I see evidence of the struggle already under way. Stands of *Acropora* corals rise in tangles as chaotic and hefty as century-old blackberry thickets. When alive, this must have been among the most spectacular staghorn coral forests on the globe. But now it is little more than a deadhouse, a bleached bone field of thousands of coral skeletons piled in the shallows. Years ago I snorkeled a remote beach in Newfoundland that had once been the scene of the annual slaughter of pilot whales, and the bones of those dead and the bones of Tuvalu's dead look eerily similar.

There are living things here, though. As with Rangiroa's Lagon Bleu, the superheated waters of Funafuti's inner lagoon form a perfect refuge for juvenile fish. The pieces of rotten seaweed rolling along the bottom are not algae at all, but the juvenile forms of dragon wrasse (*Novaculichthys taeniourus*), whimsical two-inch-long creations of green and white and yellow and brown nearly as populous in Funafuti's lagoon as are fallen leaves in the temperate woods. These cryptic youngsters are energetically acting their costume, fluttering in the surge, flipping from one tack to another, drifting backward and sideways, and bumping into ripples in the seagrass rubble.

Although most of the staghorn corals on this patch reef are dead, their bulwark remains impenetrable, and it's hard work finding a weak line through their defenses to deeper water. The lagoon here is so shallow that I find myself performing the aquatic equivalent of crawling on my belly, while the sea temperature is so high that I have to fight the part of myself that wants to flee for the merely sweltering sunlight on the beach. About one hundred feet from shore, I find a hole in the bulwark no bigger than a Volkswagen van. This provides a passageway to deeper water, to a microscopically cooler realm, where the seemingly dead reef is springing back to life. Out here, tiny colonies of baby corals decorate the tips of the lifeless staghorns like gaudy blue and pink fingernails.

Unlikely as it seems, this nursery is the frontline of the battle upon which every creature and person living in Tuvalu depends.

The colorful infant corals are capitalizing on acres of available turf suddenly devoid of established adult corals — though they compete for the vacant real estate with fast-growing algae. Their seaweeds' presence attracts hordes of browsers and grazers — herbivorous parrotfishes, surgeonfishes, rabbitfishes, blennies, damselfishes, mollusks, and sea urchins — who inadvertently scrape away the delicate polyps of the baby corals, undermining the reef's recovery.

Although capable of living a thousand years or more, coral reefs grow only incrementally, less than an inch a year. If Tuvalu's corals cannot multiply fast enough to overcome the appetites of this vegetarian army, then the reefs will be converted to sand that even smallest waves will wash away, taking the slivers of *motu* with them.

21

Liquid Faultline

GRANTED, I AM visiting Tuvalu at the height of the austral summer, during the muggy, mosquito-infested rainy season. But even so this atoll is beginning to resemble the hottest seaside place I have ever known. When I ask Rolf if there is a fan for my room, he brings me a mildew-stained desktop model — a breezy delight until later that night, when the fan's song changes from a soothing purr to a squawking bird. I try changing the speed, turning it off and on, refocusing the fan. But nothing helps. By dawn, the squawking is outcrowing the roosters, and the blades are moving through sludge. I have long since abandoned sleep for reading the small yet explosive epiphanies of Japanese tanka.

> *Come quickly — as soon as*
> *these blossoms open,*
> *they fall.*
> *This world exists*
> *as a sheen of dew on flowers.*

Later that day, Rolf, swaying with beer, on purple legs now swollen from infection, attempts to fix my breeze. But he has no WD-40, and so he lays the fan on its face in order to pour two oversized gulps of motor oil directly into the motor. I am no *enginyuure,* yet am alarmed enough to step away — although not far enough. When Rolf turns the fan on, a spray of oil covers me, my bed, and the floor around the bed.

Look, says Rolf, shrugging, it's finished. It's done. Nothing survives here.

He stumbles off and returns a few minutes later with another fan, an antique red model with no blade guard. I don't care about its lethalness, though, because when he plugs it in, it blows like a typhoon, ripping the *lavalava* right off my bed and sending the mosquitoes and the mugginess somersaulting.

Rolf celebrates the victory by sitting on my oil-soaked bed, crossing his purple legs, and opening the book I left there. The pages are crackling like fire in the draft from the fan, and he is reading through what seems a blur of oscillation. But he is turning pages, and I sense a hunger deeper than his hunger for *blutwurst*. In his own way, Rolf long ago abandoned the world for a hermitage by the sea, among people whose language he still can't speak.

I ask him if he's familiar with the Japanese concept of *yugen*, but he does not look up from the book. There is no real translation in English, although perhaps there is in German: the yearning for the sadness accompanying the loss of incomparably beautiful things. Each exposure to *yugen* grows on itself. On a sliver of land like Funafuti, the feeling is never far away because the sea brings everything and takes it away again: coconuts, unmated flip-flops, the occasional weary migrant (lizard or beetle) afloat on a mat of rotting vegetation. Living atop such tenuous ground in the presence of an overwhelming sea, *yugen* compounds exponentially, and this may be why the language of Tuvalu sounds like sighs.

When the first Polynesians arrived from the verdant, high islands to the south and east they found life on Tuvalu's atolls arduous. They became dependent on the simplest of foods: coconuts, the pigs they brought with them, the fish they could catch, and whatever threadbare crops of *pulaka,* a tarolike root, they could coax to life. When high winds and waves from tropical storms and cyclones submerged their low-lying islands, as they sometimes did, the people tied themselves to spindly coconut palms, hoping the wind might spare these fragile anchors.

Now a changing global climate promises bigger and more frequent storms in the Pacific, and more dangerous floods in the low-lying coral atolls. This is evident in the flooding caused by the King Tides, the seasonal high tides that sweep across the islands and sometimes bubble up from the soil as seawater springs. Once a scourge every February, the King Tides now occur erratically for nearly half the year, from November to March. Likewise, the big cyclones that tore through Tuvalu once or twice a decade now occur with frightening regularity. The 1990s produced seven of them, and the storm surges from these storms washed away the 125-acre *motu* of Tepuka Savalivili on the far side of Funafuti's lagoon.

Rolf points to an empty spot on the horizon. Right there, he says. It used to be right there.

We stare out the double doors of my room. Rolf describes how much fun he and Emily used to have on the *motu* years ago, when there was still a population of *paalagi* living in Funafuti. On weekends, these European residents and their South Pacific girlfriends would sail a catamaran out to Tepuka Savilivili with a picnic basket and beer, and spend the day snorkeling in the paradise that was their home.

Look, says Rolf, shrugging, it's no more.

The disappearing *motu* notwithstanding, Rolf declares that the whole business of rising sea levels is preposterous.

Och, he sputters. The islands aren't disappearing. I'm an *enginyuure*. I can see such things. Nothing is disappearing.

I wonder if he really believes this when the sensation of threat is so ever-present: the sea on both sides, the constant drumroll of surf, only a thin strip of land between — like living on a liquid fault line.

He cites the building boom under way in the village, and asks me why the governments of so many nations are investing in offices and houses in Tuvalu if the whole country is destined to sink beneath the waves. Yes, the weather is all wrong, he admits. It rains in the dry season and does not rain in the rainy season, but the disappearing *motu* of Tuvalu are irrelevant.

In fact, across Funafuti a building boom *is* transforming the atoll from its subsistence roots to a modern, noisy, polluted, and overcrowded Third World village. Only twenty years ago the island was sparsely populated with thatched *fale* and people who drove no cars on an island without roads. When Rolf and Emily arrived from Nauru, there were only three motorcycles in the whole country, and one of those was Rolf's. When they built their house, the only two-story structure in the nation, it instantly became the center of Tuvalu's social and diplomatic world, hosting functions for the prime minister and parties for the Catholic Church.

Even though I'm not Catholic, laughs Emily.

Even though I'm not a believer, says Rolf.

You do believe, says Emily.

I don't.

Now Funafuti is bursting with Tuvaluans moving here from all the outer atolls, chasing down the few government jobs available, or training at Tuvalu's Maritime Training School, which supplies merchant mariners for ships around the world. In the handful of congested blocks that make up the village, the sounds of construction peal almost as often as the church bells, with much of the noise coming from the center of town, where the three-story Government Building is under construction. Destined to be the tallest structure in the nation, a veritable high-rise by Tuvaluan standards, this building is a thank-you gift Taiwan (which won a recent round in the Pacific cold war by convincing the Tuvaluans to formally recognize Taiwan as the real China).

Nearby, the new Princess Margaret Hospital has been funded by Japan, which is trying to convince the non-whaling Tuvaluans to cast pro-whaling votes in upcoming meetings of the International Whaling Commission. At both construction sites, Fijian and Tuvaluan workers lounge in the shade, while their Australian foremen march around in ubiquitous Blundstone boots and khaki shorts. As with Rolf — who simultaneously believes that nothing survives, and yet these islands will last forever — Tuvalu appears caught in a tidal cycle of doubt, ebbing and flowing between plans to abandon the country and hopes of reviving it.

Because there are no motorscooters or motorcycles or even push bikes in working order for me to rent, I am hitchhiking here, although thumbs are not required, simply a heat-stricken gait on the side of the road. No car or truck passes without offering a ride, and it proves a good way to meet the locals, albeit only men, who immediately inquire as to my marital status. Recently a Funafutian married a *paalagi* woman, and his reports on the novelty of my kind are apparently piquing some interest.

Most of these men turn out not to be native Funafutians but transplants from Tuvalu's outer atolls, who have come here in search of economic opportunity. They avoid talk of rising seas, their conversation instead turning to more immediate concerns. The weather is all wrong. And eight months ago, this road — the only road in the country — was paved along a seven-mile length, which everyone agrees has made the island much hotter, since the black tarmac has usurped so much white sand. One elderly passenger complains that the Funafutians won't walk anywhere anymore, and worse yet, they won't go barefoot but insist on wearing flip-flops. He blames this preponderance of newfangled footwear on the road too, saying the pavement is too hot to walk on, even for coral-calloused feet.

There is little or no television here, only a few hours of radio a day, and most of these drivers and passengers have never been farther than their home islands — although some have traveled to Fiji or New Zealand, and a few, working in the merchant marine, have been all over the world. But most don't have much to compare their country to. When I mention a report issued by the U.S.

State Department that describes an apparent human-rights paradise in Tuvalu, a world devoid of killings, disappearances, torture, and refugees, as well as a nation graced with universal literacy and almost no violent crime (the only jail is currently empty), the Funafutians smile and nod politely.

But whereas I had expected to meet a nation of people eager for me to broadcast their plight to the world, instead I am finding citizens wary of the topic of sea levels. To a person, they seem quietly disappointed that I am not a tourist. Perhaps they are afraid that too much talk of flooded islands will squash any hopes of tourism ever establishing here.

Rolf tells me of a German tourist, no, French — he can't remember which — who spent a week at the Hideaway Guesthouse doing just that: hiding away. He immersed himself in books, barely leaving his room. On the morning of his departure, the truck Rolf arranged to take him to the airport never arrived. Fearful of missing the one flight a week out of the country, the visitor sprinted down the road to the Catholic church, where he begged a ride. He panicked again at the airport when the Air Marshalls turboprop engine began to rev. Fearing the plane was leaving without him, he burst past the waiting passengers, across the tarmac, up the retractable stairway, and onto the plane. An immigration official and two burly airline employees unbuckled him from his seat and dragged him back to the terminal to formally check him onto the flight and out of the country.

Tuvalu has a long way to go before tourists will find any comforts here. Yet I sense that its citizens' reluctance to talk of rising sea levels stems from more than a desire for these elusive visitors. Perhaps they are expressing a common human optimism that catastrophe cannot happen here. Or perhaps they are masking their shame, as if they feel responsible for their impending status as *fakaalofa* — the Tuvaluan word for landless, which means, literally, deserving of pity.

22

Leave Your Values at the Front Desk

THE JAPANESE FORM of poetry known as tanka was born in the Japanese Imperial Court and reached its apex among female writers. A lady-in-waiting might compose the first three lines of a tanka before sending it to her paramour, then hold her breath, so to speak, awaiting his finishing couplet.

> *Yesterday,*
> *what were my reasons*
> *for sighing?*
> *This morning,*
> *love is more painful still.*

In the science of ornithology there is a similar form of discourse between males and females known as duetting or antiphonal singing. These duets or choruses occur in one of three ways: either the males and females sing the same song together; they sing completely different songs; or they sing parts of the same song alternately yet in such perfect coordination that it sounds like the work of one bird. (Some rainforest species employ singing septets.) Whichever format is employed, duetting is considered a marital aid, strengthening the pair bond and reinforcing the auditory fenceposts of the nesting and feeding territories.

Some human partnerships develop a kind of duetting, and

Emily and Rolf have one of the more practiced repertoires I've heard. At Emily's office in Tuvalu's family-planning clinic, where I have brought Rolf because of his hugely swollen feet and legs, Emily proclaims: You need medicine.

I'm not sick, he scoffs.

Your legs are infected.

Och. It's nothing.

I'm going to give you some medicine.

I don't want any medicine.

You have to take this medicine.

Leave me alone.

Here, take this.

Let me die.

If you die, I will wrap you in a pandanus mat and let the dogs drag you into the sea.

I don't care. You can put a sign on me saying, *I'm free.*

On the short walk from Emily's clinic to the bank, Rolf is so weak and lags so far behind us that we have to make frequent stops. Emily tells me how he always hated it when she walked behind him when they traveled in Europe, and how he always made her hold his hand. Just look at him now, she whispers. Behind us, Rolf is sputtering in outrage. I like walking back here, he counters. I get to look at your backsides this way.

A third voice is sometimes added to Rolf and Emily's duet, that of Father Kamilo. The Emily-Rolf-Kamilo triad goes back sixteen years, when the Father arrived in Funafuti from a thirty-year posting in Samoa. Now seventy-four years old, he has been in the South Pacific since 1956, a far cry from his cold roots in Quebec.

Both Emily and Rolf use the Father as a kind of marital sounding board, and when Emily asks about Rolf's drinking, Father Kamilo says that God (pointing heavenward) could make Rolf give it up like that if he wanted (snapping his fingers). When, for the umpteenth time, Rolf finishes his habitual dependent clause (after thirty years of marriage) with one of his many modifiers (she bosses me around . . . she won't talk to me), Father Kamilo chimes in that if God (pointing heavenward) wanted things to be different, he could make it so just like *that* (snap).

When Rolf had worried about what to do with me my first morning in Tuvalu, he delivered me into the care of Father Kamilo. I could see that I was not the first guest brought his way, and he greeted me courteously but cautiously. As I soon discovered, however, the Father is a cheerful soul, with a great love of jokes, and the ice between us broke as he began to feed me pages of jollity sent to him from friends around the world.

One of his favorites is a series of English mistranslations called "And Now For Something on the Lighter Side":

> *In a Bucharest hotel lobby:* The lift is being fixed for the next day. During that time we regret that you will be unbearable.
> *In a Leipzig elevator:* Do not enter the lift backwards, and only when lit up.
> *In a Paris hotel elevator:* Please leave your values at the front desk.
> *In a Zurich hotel:* Because of the impropriety of entertaining guests of the opposite sex in the bedroom, it is suggested that the lobby be used for this purpose.
> *On the faucet in a Finnish washroom:* To stop the drip, turn cock to right.

The Father feeds me other drolleries whenever we meet, including pages from the *Marist Messenger* of *Reader's Digest*–like jokes called "Can You Bear It?" Humor is his ally in the foxhole of this posting, where he claims, only half-jokingly, to have been forgotten by his superiors, although if God wanted . . . (snap).

He speaks longingly of Samoa, its beautiful vistas and varied cuisine, and relates how he was sent from that paradise into what I gather feels like his propitiatory exile on the slim evidence of fifty dutiful Catholics subsisting without a shepherd on Funafuti. When he arrived, however, only twenty-six of these faithful were to be found, the number having been fleshed out with the dead to meet the minimum required for a new parish.

Since then, the number of live Tuvaluan Catholics has grown to one hundred sixteen — an average of five or six a year, I calculate, probably the same as the Catholic birthrate. The Father hems when I inquire as to whether he's actually persuaded any of Tuvalu's staunch Protestants to leave the century-old clutch of the London Missionary Society. When I make a suggestion as to why this might be — with a snap of my fingers — he stares at me in disbelief, then throws back his head and laughs.

Because I am on foot in Funafuti, moving slowly through the heat and the afternoon thundershowers, I have ample time to ponder the meandering truths of this place. In 2001, Tuvalu began actively lobbying Australia and New Zealand to accept its entire population as environmental refugees in the event the nation succumbs to the waves. Australia refused, citing its tidal gauge data — although this was more likely a cover for Australia's strict no-refugee policy. New Zealand agreed to accept Tuvalu's citizens, although only eighty a year, and only for three-year periods.

Meanwhile, Tuvalu's atolls will likely become uninhabitable long before the point of total submersion. Floods and rogue waves raise the saltwater table underlying the *motu*, poisoning the Tuvaluans' staple crops. Already some farmers have been forced to grow their *pulaka* in metal containers, and already some of the smaller *motu* have lost their coconut palms to saltwater intrusion. Nor are storms a prerequisite for disaster. "Last August," Prime Minister Saufatu Sopoanga told me, "on a clear, calm day, a sudden wave surge rolled in from the sea and washed across Funafuti into the lagoon, flooding houses and *fale*." There was no apparent reason for it.

Yet, paradoxically, Funafuti is building like a nation with a long-term future. Along with the new public buildings, there are scores of new houses going up, some grander than the once-incomparable Hideaway Guesthouse. Presumably in acknowledgment of the rising waters, these new homes are being built on ten-foot-tall stilts, notably different from the traditional dwellings. Of course the stilts might also be a response to the mounting tide of garbage on the atoll.

Until recently, the only refuse the Tuvaluans created was biodegradable coconut husks and fish bones. In keeping with past practices, they continue to toss their trash more or less out their front doors. But now much of this new trash clogs the borrow pits left over from construction of the World War II runway, when American troops excavated fourteen million cubic feet of precious atoll real estate, creating holes the size of Olympic swimming pools.

Today these pits are located in the most congested neighborhoods on Funafuti, where they flood with each rainfall, and each King Tide. Wet or dry, they overflow with floating coconut husks, beer cans, disposable diapers, plastic soft drink bottles, shampoo bottles, detergent bottles, baby formula bottles, cigarette butts, cigarette boxes, flip-flops, car parts, motorbike parts, potato chip bags, vacuum-packed milk cartons, and cans of every kind of food, including canned coconut milk.

Funafuti has become too crowded for its inhabitants to grow their own food, as they did for two thousand years. Today the Funafutians must buy their sustenance from Australia and New Zealand, complete with the Western world's excessive packaging. Children and piglets and puppies play on the edges of the borrow pits where all this packaging festers, the dumps garnished with human, hog, and dog manure. These fetid, half-submerged wastelands may be the most chilling preview of Tuvalu's drowning future.

23
Little Cemeteries

L UNCHING ON THE OUTDOOR terrace of the Vaiaku Lagi Hotel, I encounter Tuvalu's tiny population of expatriots, the ten or so mostly Australians
and New Zealanders, the majority of whom are aid workers of
some kind. Most are as suspicious of me as a crocodile on their
picnic blanket. Still, I ask to sit with them, joining their midday
scene on the shimmery edge of a lagoon filled with half-sunken
fishing dinghies and floating litter. This gathering is a daily event,
the drinking of Victoria Bitter, the consumption of Tuvalu's luxury foods — slices of canned meats in Wonderbreadlike sandwiches colored magenta from the relish of canned beets.

A former AusAID worker, Paul, opens up when I ask about the
garbage situation on the atoll, and everyone at the table chimes
in to say that the present scenario is a huge improvement from a
few years ago, when the borrow pits were literally borrow mountains of refuse. Paul tells of his past work implementing a system
of garbage collection and garbage sorting, separating the organic
waste from the inorganic. The alkaline soils here, he says, can accept all the plastics and metals without leaching their dioxins and
heavy metals, as long as they're kept separate from the organic
waste.

In keeping with everyone else on this island, the expats at the
table scoff at the notion of disappearing islands. Paul jokes that

sea levels may or may not be rising, but for sure the housewives of Tuvalu are sweeping the island away each morning. When garbage collection began, he says, 70 percent of it was organic waste from those sweepings.

His story triggers a voluble discourse at the table. Everyone has tales to tell of the odd Tuvaluan preference for dirt yards, and how the women crawl on their hands and knees plucking even the tiniest green seedlings from the earth around their homes. The aid workers laugh and shake their heads in shared cultural confusion — and I realize that these denizens of the terrace of the Vaiaku Lagi Hotel, with its air-conditioned bar and its imported beer and boiled meats, are also resisting contact between the organic and the inorganic, holding fast to their own little poisons.

Perhaps the Tuvaluans' method of yard management is based on a biological or epidemiological reality we don't understand. At any rate, this custom is one of the things I like best about this place, and one I fear will disappear upon abandonment. Transplanted to New Zealand, the people of Tuvalu will doubtless learn to grow lawns.

Even conservative forecasts now predict that the earth will warm three-and-a-half to ten degrees Fahrenheit over the next century. Spread over decades, a three-and-a-half-degree rise is akin to moving the climate bands poleward thirty feet a day — and although land animals might be able to move to keep up with that kind of shift, islands cannot. Even worse, as the climate bands shift northward in the northern hemisphere and southward in the southern hemisphere, the melting of the icebound regions sets in motion a feedback loop difficult to escape.

This mechanism is regulated by the earth's albedo (Latin *albus*: white), the fraction, measured between zero and one, of light or heat reflected by a surface. A white body has a high albedo. Virgin snow, for instance, rates a 0.95 albedo. Compact sea ice, with an albedo of 0.80, reflects 80 percent of the sun's heat back into space, while seawater, with an albedo of only 0.20, absorbs 80 percent. Consequently, any reduction in the ratio of ice to water further increases the warming of the oceans. As the seas warm, they rise further from thermal expansion, creating an even greater surface area of water, which promotes further warming, and further melting. "Once the process is set in motion," warns Robert Watson, the chairman of the Intergovernmental Panel on Climate Change, "it cannot be slowed down in anything less than a few millennia."

The Tuvaluans face a difficult choice. If the seas rise and they stay in Tuvalu they will die. But if they leave, some part of them will die. In the event of abandonment, Prime Minister Sopoanga says, "we would like to stay as close to Tuvalu as possible, where we could still have the same water and the same air." Despite a prevalent Western belief that all the world desires to immigrate to its shores, the Tuvaluans feel differently, and the government of Tuvalu is currently buying land in neighboring Fiji in the hope of staying close to home.

Father Kamilo says that although many of his parishioners mistake New Zealand for heaven (pointing upward), few actually wish to leave the nation they pledge their allegiance to as *Tuvalu mo te Atua* (Tuvalu for God). Catholic or Protestant, most Tuvaluans hold firm to the Genesis story, with its promise that rainbows are proof of God's assurance that, after Noah and the Ark, he will not flood the earth again. Buoying this optimism every day, from every quadrant of Tuvalu's skies, neon-bright rainbows blossom and fade.

Father Kamilo also disdains the disappearing island theory, citing the contradiction of the building boom. Yet the news of even the nearby world could easily pass him by. At his desk in his tiny office, opening his mail, he informs me that letters from England arrive "pretty fast," as he opens a Christmas card on Valentine's Day.

Despite his seventy-four years, Father Kamilo is as fit and athletic as a man in his forties. In fact, he seems half the age of Rolf, who is sixty, or something like that according to Emily, or sixty-two, sixty-one, or sixty-three on the occasions I ask Rolf himself. The Father attributes his health to a contraption called the MegaMag, an electromagnetic devise with ring-shaped and rectangular-shaped attachments that look hefty enough to treat horses, which he purchased on a visit to Canada years ago. He swears that electromagnetic treatments applied to his skin with the MegaMag are keeping him alive. And apparently they're keeping other Tuvaluans alive, too. The Father shows me his photo gallery of people treated for tumors, burns, broken bones, paralysis, and all the other afflictions that suddenly bewilder the healthy. He is straightforward with his clinical synopses: this one recovered; this one died. He tried the MegaMag on Rolf, but admits it can't cure everything.

Somewhere in the Father's mind must be the realization that, electromagnetics notwithstanding, he will likely die on Funafuti, having possibly never converted anyone, having never been posted back to the bigger world, and having never seen any of the outer atolls of Tuvalu. Perhaps he will finish up like all the other Tuvaluans interred in the private cemeteries gracing the front yards of most of the homes here, what the Father calls the dead centers.

The Hideaway Guesthouse has just such a cemetery, where Emily and Rolf's eleven-year-old son is buried. This, I suspect, is the source of Rolf's despair. Clearly there was a time when he cared enough about life to build the lodge, complete with features that don't exist or are uncommon elsewhere in Tuvalu: a second story, window screens, screen doors, tiled walkways, porches, a screened-in patio. Although all is in disrepair these days, there was a time when he mended his world, despite the fact that all he talks about now is how hopeless it all is, how his car died (the salt air), and his motorcycle died (the rust), and his camera and his stereo died.

Like most graves here, Rolf's son's inhabits a small rectangular plot segregated from the rest of the front yard by hogwire fencing decorated with plastic flowers. Some graves are paved with upside-down Coke bottles left over from World War II, planted in the earth in patterns of green, brown, and clear glass. Most graves are roofed with corrugated aluminum. Rolf says the Tuvaluans always provide shade in hopes the dead will sleep well, adding, for the hundredth time: It's such a strange place, they're such strange people. His son was initially treated for leukemia in the clinic in Funafuti, and when that failed, Rolf took him to a hospital in Suva, Fiji, and when that failed, he took him on the long journey that the Polynesians of old knew well, to Aotearoa, where his son died.

I came back to Tuvalu with a coffin, he says.

Another time, Emily tells me that there was no treatment, nothing could be done. She says his death "wrecked" her for a long time. Silently, I calculate backward, realizing that her son's entire short life spanned the era of French nuclear testing, when radioactivity blanketed the South Pacific, spawning many cases of leukemia. I also realize that sixteen years ago there was a treatment about as good as a cure for many childhood leukemias, only not available in Tuvalu, apparently. But I don't want to tell her that, so I say that treatment is available now, a very good one.

Oh, she says, with a mixture of hope and pain.

It occurs to me that after two thousand years of human habitation, a fair amount of Tuvalu's tiny landmass must be composed of the bones of its people. When I think of the future, this thought saddens me: what will become of these other Tuvaluans, the ones whom the people still consider important enough to erect roofs over their graves for shade? Surely the New Zealanders will not accept the dead Tuvaluans or the soil they have become.

It rains most afternoons on Funafuti, sudden violent squalls complete with lightning and cloudbursts that funnel water along the atoll's one road before emptying, creeklike, into the wind-torn lagoon. Joining the shouts of thunder are the Tuvaluan churchbells pealing wildly in many keys from all directions. I try to be home from the village before this daily deluge, and spend the duration of it in the screened porch attached to my room, enjoying the periods of gusty winds and the shivery hula of the palm fronds. Between squalls, the air swells as thick as steam and the mosquitoes swarm thickly on it. Yet I am spared their assaults, thanks to the red fan, devoid of blade guards and all the more powerful for it.

Most afternoons, Foliki (Tuvaluan *foliki:* small), the household's five-month-old orange kitten, joins me. Foliki was abruptly exiled from her mother's affections after the birth of a new litter (which none of us has yet seen), and is lonely for any company, even *paalagi* company. Both mother and marmalade kitten are the smallest cats I have ever seen, skinny and hungry as starvelings. I raid the kitchen each afternoon for whatever I can find — congealed lamb stew, stale bread, fish tails, coconut pudding — which they eat voraciously and without prejudice, although the kitten begs hopefully for its mother's milk.

The rains usually end in the last hour of daylight, leaving the atoll a modicum cooler and fresher than before they arrived, and polishing the pewter canvas of the distant lagoon. The nearshore lagoon, however, wears the perturbations of the squalls for another hour or so, its lapping waves littered with palm fronds and mangrove leaves, its shallows muddied with soil runoff from the land. A few newly launched plastic bottles bob along the shoreline, destined to be grounded again when the tide retreats.

After the rains, a handful of kids race out to romp in the shallows, though, compared to other parts of Polynesia, the people of Funafuti seem remarkably landlubberly. I see only one *vaka* and no surfers at all. The adult swimmers frequent the water only in the early mornings and late evenings to make use of the lagoon's plumbing — what Father Kamilo calls the vicious cycle (people defecating in the sea, fish eating feces, people eating fish). Of course, before five thousand people settled on this atoll, such a practice would not have been any more problematic than fish droppings in the water.

Latrine or not, I snorkel most evenings after the rains, winding through the labyrinth of the dead staghorn corals, looking for a route to deeper water. Eventually I locate a line to the backside of the lagoon pinnacles, at the edge of the clear outer lagoon, where, thanks to a regular tidal flushing from the open sea, the water is cool and clear. Here, at last, is a spectacle of vigorously recovering corals, the baby colonies repopulating the white skeletons, and the water column rich with a colorful contingent of herbivores, corallivores, piscivores, filter feeders, benthic feeders (Greek *benthos*, depth of the sea), and planktivores. The new corals adhere like skin grafts on a badly burned patient, patchily distributed, not yet unequivocally accepted, but presenting an altogether more optimistic picture than the inner lagoon.

Better yet, the outer lagoon harbors genuinely ancient and still-living corals, including some healthy thickets of staghorns unrivaled by anything I've ever seen, along with plate corals big enough for a banquet. The density of life here is exponentially greater than anything in the inner lagoon. Biodiversity oozes from the confines of the reef, as if an aquarist had grossly overstocked the fish tank. Threaded deep into the tangled arms of the corals are schools of bluestreak cardinalfish (*Apogon leptacanthus*), as clear as glass except for the blue and yellow pattern around the gills and into the eyes. They lie inside the shelter of the staghorns, sleeping on their sides, their backs, their heads, or their tails, awaiting night, when they will emerge into the water column to feed on plankton.

Already plucking zooplankton from the far edge of the coral outcropping are dense schools of squarespot anthias (*Pseudanthias pleurotaenia*). The females wear a brilliant goldfish orange with faint purple stripes the lengths of their bellies. Clouds of them rise and fall with the surge from the surface, their numbers nearly masking a single male accompanying them. He is larger than the females, with elongated fins and a square-shaped magenta spot on his flanks that glows with unearthly radiance, as if lit by a black light. At some point, he was a female too, but changed gender when the job fell vacant.

With the sun plummeting towards night, I paddle back to a nearly empty shore. The splashing kids are replaced by a few adults performing their end-of-day ablutions. They are far away and engaged in things Rolf asked me, please, not to write about. Yet I feel an urge to swim over and tell them the good news from Funafuti's outer lagoon.

24

Diving the Apocalypse

E VEN IF FOR SOME miraculous, rainbow-tinged reason sea levels do not rise in the coming years, the islands of Tuvalu are still at threat from the warming sea temperatures associated with the El Niño-La Niña/Southern Oscillation. Until recently, Tuvalu was a cyclone nursery of the Pacific Ocean, where baby low-pressure systems were born before moving away to decimate distant shores. But with the onset of more frequent and more powerful El Niño events, Tuvalu has found itself the victim of fully fledged cyclones. Even without storms, El Niños kill reef-building corals, which are extremely sensitive to temperature changes and can survive only within a tiny margin of eighty-six degrees Fahrenheit. When exposed to higher temperatures, the coral polyps expel their zooxanthellae partners — the source of their brilliant colors — rendering the reef a ghostly white.

Bleached corals soon begin to suffer from a lack of proteins, lipids, and carbohydrates, a condition that undermines the polyps' skeleton-building and reproductive cycles. The animal part of the coral may survive a few days like this, but the divorce from their plant partners eventually leads to necrosis of their tissues and death. The dead reef is soon invaded by algae, and thereafter by those who shelter in its skeleton, or those who graze on the plantlife, the bioerosive army that exceeds the rate of carbonate production with such efficiency that the reef is quickly converted to sediment.

The forests of dead staghorn corals in Funafuti's inner lagoon are likely victims of recent El Niño warmings. The venerable corals still thriving in Funafuti's outer lagoon were likely spared the worst effects because they inhabit cooler waters. Or perhaps they bleached and then recovered. Recent studies show that the long-term effects of coral bleaching can be minimized if the temperature spikes last only a few hours or a few days. In the wake of such ephemeral events, the corals invite the free-floating zooxanthellae back aboard as soon as sea temperatures moderate.

Opportunity is inherent in this periodic divorce between polyp and plant. Andrew Baker and his team from New York's Wildlife Conservation Society found that some polyps repartner with zooxanthellae species adapted to higher temperatures, enabling the animal/plant symbiosis to adapt to changing sea conditions. This news offers genuine hope for the future of coral reefs — though it's tempered with the reality that in the past two decades, worldwide coral bleaching events associated with El Niños have destroyed reefs across entire ocean basins.

A 2004 report by the Global Coral Reef Monitoring Network warned that global warming is the single greatest threat to corals, with 20 percent of the world's reefs so badly damaged that they are unlikely to recover, and another 50 percent teetering on the edge. The report predicts that within the next fifty years massive bleaching events on the order of the 1998 El Niño, which damaged or destroyed 16 percent of the world's reefs, will become a regular, possibly annual, occurrence.

Meanwhile, a computer simulation from the National Center for Atmospheric Research in Boulder, Colorado, suggests that global warming, not a meteor strike, triggered the earth's most severe extinction event. The Permian-Triassic extinction, also known as the Great Dying, came close to destroying all life on earth. New data suggest that an enormous rise in atmospheric carbon dioxide two hundred fifty million years ago — possibly fueled by massive, earth-building volcanic eruptions in Siberia — led to a global warming of about eighty-six degrees Fahrenheit. This precipitated the extinction of 90 percent of all marine species and 70 percent of all terrestrial vertebrates, leaving fungi to rule the world for many an eon.

There is no sand beach on the ocean edge of Funafuti Atoll, only a coral shore composed of chunks of broken staghorn, elkhorn, brain, fan, lettuce, and plate corals, most the size of golf balls or baseballs, some as big as basketballs, and a few as big as car tires. These coral rocks are so white, dry, and porous that the atoll appears to be made of bones.

For the most part, the seawater washing onto the shoreline arrives gently on this skeleton beach. The swells rising and curling from the back of the ocean are not breaking here but thirty feet offshore, on the berm known as the algal ridge. This is the most violent zone in the reef community, the marine equivalent of a bombing range, where combers pound ceaselessly day and night. While no corals are hardy enough to survive such an assault, a variety of marine plants thrive on it.

These are the encrusting red algae, or coralline algae, marine plants that grow as hard as rock. Many are brilliantly colored, appearing as red, pink, and purple paint splashed across the reef. These plants remove calcium from the water and produce calcium carbonate, a cement they use to bind the abundance of loose rubble on the reef into a stable mass. In this manner, broken corals, broken barnacles, limpets, seashells, sea urchin spines, the spicules of sponges, and the calcium components of dead algae are bound by the matrix of coralline algae into a mix so stalwart that the living algal ridge resists the erosive force of crashing waves.

In fact, the overall structure we call the coral reef is created as much by the efforts of these plants as by the corals. Coralline algae secrete at least half the limestone laid down on the reef each year, forging the I-beams, studs, and joists of the house of coral. An equally valid name for these architectural wonders in tropical seas would be algal reefs. And like corals, the coralline algae are also vulnerable to increasing atmospheric concentrations of carbon dioxide absorbed into the oceans and intensifying the acidity of seawater, thereby slowing the calcification process of both corals and coralline algae.

Increasingly acidic oceans force reefs to grow more slowly or not at all, and some scientists predict that by the time carbon dioxide levels have doubled, around 2050, reef construction will be reduced to 40 percent of its present efficiency. Fewer reefs will, in turn, undermine the ocean's ability to absorb and mitigate carbon dioxide in the atmosphere, leaving those of us in the dry world more susceptible to the cascading effects of global climate change, whether we live on fragile islands or on solid continents.

In the wake of the afternoon thundershowers, as Funafuti's children romp in the waves, I wander to the edge of the lagoon to place phone calls home. My satellite phone requires clear skies, and, for reasons I can't fathom, reception is best while I'm standing in the lagoon, the antenna on the phone raised, and my head and upper body contorted into strange, birdlike poses in order to maintain contact with the signal. Even so, contact is frequently broken, and so I place my calls, on average, four or five times per conversation, striking a new pose each time. Cordless landline telephones have not yet arrived in Tuvalu, and cellular phones are unlikely to get here anytime soon. What I'm doing must seem all the more incomprehensible to the kids in the water, who observe and occasionally mimic my every move.

Funafuti is most beautiful in this hour after the thunderstorms. From where I stand in the lagoon, with the borrow pits and the garbage piles behind me, I can admire the loveliness of an atoll that not many years ago would have been lovely in all directions. To the west, the water is a pearly green, and the sky is filled with layers of clouds in every color of gray, white, silver, and pewter, winking blue windows. At this hour, facing west, it's possible even for a *paalagi* to feel nostalgia here.

During an oft-interrupted phone call, a friend suggests that Tuvalu might have a bright future as a postapocalyptic tourist destination, where a visitor could lodge on the second floor of the hotel while scuba diving its first floor. I find myself assessing the potential of future attractions on the atoll. The lobby of the Vaiaku Lagi Hotel would make a pleasant dive site, open and airy (watery), with the guest rooms adding an exploratory thrill. All would be clothed in pretty corals, if sea temperatures permit. The windowless kitchen would provide an excellent daytime sleeping site for whitetip reef sharks, the small dining room could house a large napoléon, with ample room behind the bar for a moray eel, and plenty of empty whiskey and rum bottles to make homes for octopuses. The outdoor terrace, currently home to expats, would make an ideal habitat for nervous and secretive garden eels.

I also imagine the dive potential of the Hideaway Guesthouse, calculating how many coral heads would fit in the kitchen, and where the parrotfish might sleep. This evening, sipping tea after a dinner of curried flying fish, Emily proffers that she doesn't know whether sea levels are rising, or whether Tuvalu will disappear beneath the waves. But she is worried. She asks me what I think, as if I have some secret information. Whispering, trying not to incite Rolf, she confesses that she doesn't want to leave here, no matter what. I think I understand, and imagine many Tuvaluans who might tie themselves to coconut palms rather than sail away forever.

Rolf is slumped on the sofa, hiccupping and snoring above the mewing of the tiny kittens the mother cat moved under the sofa earlier today. He has worn himself out this evening with story-telling — a story I have already heard, but never mind, Emily wasn't here to listen the first time, and it's really for her benefit that he tells it tonight. He winds himself into the tale of a visit from Emily's family. One of her aunts ate — Do you know how many? no, I don't — she ate *ten* flying fish. He repeats this number with so much energy that he spits.

Emily is clearly offended: That's not true.

It is true. You know it.

She was hungry, protests Emily. She was very hungry.

Och, I don't care how hungry she was. I could never eat *ten flying fish* even if I was starving.

So what? So what if she ate ten flying fish?

This is why they get fat, he turns and says to me. Because they eat *so much food*. Sometimes, they eat for days and days without stopping.

That's only for special occasions, says Emily.

It doesn't matter about special occasions. I could never eat *ten flying fish*.

She didn't eat ten flying fish.

She did and you know it, says Rolf, popping the top of his (probably) tenth beer of the day.

When he begins to snore, Emily complains that he always criticizes her people — and this after her people have gone out of their way to welcome him to their world. They treat him like a king, she says. Emily herself sometimes affectionately calls him the King of Tuvalu or the King of Funafuti, and I have seen how the whole island treats him with the fondness and respect of a doddering uncle.

He was a working-hard man, Emily frequently says, adding, but now he loves his beer too much.

When they built their home on Funafuti years ago it was a novelty, a two-story *paalagi* house way out in the bush, she says. Emily's family was mystified as to why she wanted to live so high up or so far away. Rolf built it himself and insisted on moving in before the house was completed. They would sleep upstairs, he declared, even when upstairs was only a platform without walls. That decision led to him stepping off the unfinished second floor in the dark one night and tangling with stakes of rebar. Emily managed to haul him back upstairs, but when she saw how badly his leg was broken, she ran out into the bush crying for help and flagged down someone on a push bike who rode for Father Kamilo. Rolf spent the next nine months in Funafuti's clinic, fighting the doctor's desire to amputate.

Emily moved in with him, nursing him through the long period of medical indecision, through the surgery to implant a steel bar in his leg, and through his slow recovery. Rolf has always amazed her, she says, but never more so than nine months later, when he insisted on returning to their home so high up and so far out in the bush. Hobbling on crutches, he stubbornly resumed building it.

We watch him now, snoring, hiccupping on the sofa. He's so old, says Emily sadly. He's so sick. Or maybe it's the beer.

When he awakens from this nap, Rolf resumes the evening's entertainment, regaling us with stories of his early days in Tasmania, where he studied judo until he could jump from up high and land with the shock absorbers of his arms, all of which is nearly impossible to believe in light of his present frailty. Emily and I laugh, because his stories are funny, and because they lie so close to the cusp of tragedy.

25
Nuptials

WHEN SEA LEVELS RISE, as they surely will some day, the atolls of Tuvalu will cease to manifest in the way that we know them today. There may be a brief spark in time, geologically speaking, when the Vaiaku Lagi Hotel will present as a dive site complete with resident garden eels and a big napoléon. But within the passage of one or two cyclonic storms the waves will batter it to rubble, and within one or two years afterward the rubble will be covered with the grafts of new corals — or new algae or new bryozoans or whatever is growing in the sea at that time. The *motu* formed in the last stages of a volcano's existence, the manifestation of the ancient Funafuti Island, will dissolve in the ocean currents and reconstitute elsewhere as sandbars, or as a shoreline. These new shoals will still be home, only to different beings, or perhaps to different manifestations of the same beings likewise transformed.

In preparation for whatever changes lie ahead, many young Tuvaluans are already going away, with promising students enrolling in universities in Fiji, New Zealand, or Australia. At any given time, about seven hundred fifty Tuvaluans, one in four of the adult male population, are employed somewhere at sea as merchant mariners. Yet even after these young people return, despite being richer or better educated or both, they still have no pigs, a condition considered pitiable by the older generation — though they now have the means to buy CD and DVD players and chrome hubcaps for their very own cars.

Pigs and land have traditionally been the measure of wealth in Tuvalu. Emily says Tuvaluans love their pigs and cry when they have to kill them, though when two rogue pigs rampage through her garden one morning, she doesn't hesitate to tell me that Tuvaluans are required by law to keep their pigs penned. This pair, illegally rooting through her backyard jungle, are hers for the killing.

Which would give us all a good excuse for a feast, Emily says, eyes twinkling.

Apparently, slaughtering your neighbor's pigs sidesteps the sadness issue, which gives me some insight into the Tuvaluans' past practice of cannibalism; why not eat your enemies if you love your pigs?

In fact, pigs, or rather piglets, have the run of Funafuti. While their mothers, the sows, are caged in small, roofed pens that look remarkably like the Tuvaluan graves minus the plastic flowers, the piglets squeeze through the bars and roam at will. Then comes the moment of truth, when the youngsters have acquired too much girth and can no longer squeeze into the pen, and spend their time pressed miserably against the bars of their unobtainable captivity.

Emily is a modern Tuvaluan woman, perhaps the most modern woman in all of Tuvalu, and she no longer keeps pigs, only *paalagi* houseguests. In her work as executive director of Tuvalu's family-planning clinic, she is regularly jetted around the world by the United Nations and other lofty international bodies. Yet she is deeply connected to all the families on Funafuti who do keep pigs — and because she is the groom's aunt, and because I am her guest, I am invited to the weekend wedding that has drawn Tuvaluans here from all the outer atolls to fill the Vaiaku Lagi Hotel. Father Kamilo warns me that my attendance will require hours and hours of sitting on the floor, the reason he is not attending. And Rolf tells me he is abstaining because, Och, they'll just *eat, eat, eat.* It will be nothing but food, food, food, food.

My invitation is hand-delivered to the Hideaway Guesthouse two nights before the wedding by a teenage girl and her little brother, the two of them simultaneously staring and giggling, emboldened with curiosity. The wedding invitation is a hand-made affair complete with a color Xerox of a photograph of the young couple surrounded by a hazy cloud presumably representing love. They are being married at one of the many churches of the Christian Church of Tuvalu (CCT) at ten Saturday morning.

The CCT is the offspring of the same London Missionary Society that first landed in Tahiti in 1796 and swept through the South Pacific with prudish virulence. Its doctrines continue to control Polynesian life in Tuvalu to such an extent that, despite the tropical heat, the young women here wear ankle-length skirts or dresses — not quite the Hawaiian missionary dress of old, but almost. Traces of rebellion manifest discreetly, as far from the head as possible, as if the young women could look down and say, *Well, imagine that!* at the sight of their feet fashionably shod in platform sandals.

Emily lends me a dress that is long enough and big enough for several of me and all but covers the only footwear I have, a battered pair of Tevas. What with the welts of mosquito bites, flea bites from Foliki, coral scrapes, and a *fou* (flowered head wreath) so big that it drapes over my forehead, and hangs longer than my own short hair, requiring hitching behind my ears, this may be the most unattractive moment of my life. I ask Emily to take a picture.

We primp at Emily's family home in the village. Or rather I watch the others primp, as every one of the dozen or more female relatives partakes of the communal bottle of perfume and the communal tube of lipstick lying on the table in the lounge. Only Emily's mother refrains. An ancient woman, she inhabits the pandanus-thatched shade of the outdoor cookhouse, a raised wooden platform spread with woven mats. Sitting on the floor on wide haunches, legs open in a V, back ramrod straight, she rhythmically bends and grates the *pulaka*.

I have been trying to persuade myself that my yoga practice is ample preparation for hours and hours of sitting on the floor. But seeing Emily's mother, I am not so sure, and apparently neither is she. With venerable authority she calls to a minion of the great-grandchild generation and orders him to bring me a chair. I'm fine, I say, even though she doesn't speak English. Emily refrains from translating, countering, My mother always gives *paalagi* chairs to sit on. It's her way of showing respect.

So I sit, hitching the droopy *fou* behind my ears and scraping its ant inhabitants off my cheeks. Everyone else is on the floor, and I am self-consciously raised above them at a moment when I am not feeling fit to represent my entire *paalagi* race, whatever that may be.

The thirteenth-century Zen master Dágen Zenji wrote, Now, to it! Head up, eyes straight, ears in line with shoulders. Of the many lessons Tuvalu has to offer, the lesson of sitting is one of the most instructive. This tutorial is presented in church, where the Tuvaluans — who sit as beautifully straight as a yogi in *dandasana* (staff pose) on the floor, or on the ground — instantly adopt the bad posture of bored high school students. They slump and wriggle and put their feet up on the pews and lean on their elbows on the backs of their neighbors' pews. They talk and laugh and call out to one another as members of the congregation come and go and come again, wedding notwithstanding.

Somewhere up front a minister is droning in Tuvaluan, though he can hardly be heard over the turbulence of scores of hand fans swishing through the air, and above the friendly conversations all around. Three rows ahead of me, a pew filled with children of many ages, whose heads are visible above the top of the bench, writhes like a gorgon. Somewhere up front the minister's wife is leading the congregation in periodic song, and this is the only time the congregants pay attention. Wedged between Emily on one side, singing with gusto, and her large cousin on the other side, singing with gusto, I am caught in the swaying wave between their girth. In fact, I am caught in the wave between everyone's girth, becoming one with the religious peristalsis squeezing us closer to heaven. Meanwhile, the singing and swaying hypnotically quell even the many-headed monster three rows ahead, as the voices of the adults are joined by the angelic voices of the children, all sounding incongruously light and ethereal. The men on the left side of the aisle sing in a round with the women and children on the right side.

Infused with two thousand years of cultural practice, this singing far predates the Christian influence. You might imagine a cloistered community of monks who have passed on their choral tradition from one generation to the next over the course of eighty generations. Tuvaluan song suggests the results — the effortless, intricate harmonies, the complicated singing in rounds, the perfect pitch from every man, woman, and child in the pews. There are no singing lessons here, no formal choir. Their song carries them to a new world, just as *vaka* once bore them toward islands newly risen from the waves.

The bride and groom do not kiss at the conclusion of this wedding ceremony. Instead, midway through the service, triggered by some unheard ministerial words, the groom lifts the bride's veil and tenderly kisses her on the forehead. She casts her glance downward, toward her fashionably shod feet, while the bells of churches across the *motu* ring.

At the end of the ceremony, the churchgoers abandon the discomfort of sitting upright in the pews to shuffle in a receiving line toward the newlyweds. The young couple greet their guests with formal smiles and demure handshakes, standing taut in their unfamiliar wedding clothes — the bride in full-length, full-strength puffy white Western dress, the groom in a tuxedo. Outside, a phalanx of young women wearing red-and-yellow *lavalava* awaits them, bearing a white cotton canopy on bamboo poles decorated with balloons. This pseudopalanquin is devoid of seats and requires the bride and groom to ambulate on their own feet, but it offers shade and a processional ambience as it leads the wedding party on a winding route through the village.

Parading through Funafuti's sandy streets, with barefoot neighborhood kids jumping alongside, slapping at the palanquin's balloons, the wedding party winds itself through the back streets until it arrives at a *falekaupule* (community hall) anchoring one of Funafuti's crowded neighborhoods. This is a traditional building composed of a roof on pillars and no walls, designed to block the sun but not the breeze. The bride and groom are escorted inside to a sofa draped in a white crocheted blanket, but they quickly abandon the furniture for the floor, where the bride's white skirts spread like sea foam.

Surprisingly, there are many chairs in the *falekaupule*, enough for every guest, even Father Kamilo if he were here. And the banquet is not food, food, food, food, as Rolf feared, but what appears to be simply a lot of food, albeit all of it a strange concoction of Western recipes more or less butchered by unfamiliarity and scarce resources. There are odd pizzalike things and dishes resembling spaghetti and meatballs and Wonderbread sandwiches and what seems to be canned tunafish and egg salads and many other vaguely recognizable dishes, including cakes and tarts. The guests are not, despite Rolf's prediction, eating and eating and eating. In fact they appear restrained and even listless, picking at their plates, or sipping from cans of Victoria Bitter or Fanta, while listening politely to speakers taking turns with the karaoke microphone to toast the newlyweds.

Two hours later, as the party breaks up, Emily and brethren lead me back to the church and to the really big *falekaupule* next door. The Western-style festivities we just departed, complete with a white gown and chairs and pizzas and microphones, were only an adjunct to the real celebration. Here in the giant *falekaupule* is what Father Kamilo and Rolf were talking about: a hall the size of Jerusalem decorated with balloons, streamers, plastic flowers, paper flowers, and real flowers, utterly bereft of chairs, but with pandanus mats carpeting the floor. Two platforms, one-foot-high by eight-foot-square, occupy one end of the room, each draped with colorful *lavalava* and pandanus leaves.

The bride's family is stocking the platform on the left with food. A throng of her female relatives bears coconuts and oranges and grapes and guavas and passionfruit and grapefruit and watermelons and oversized bunches of literally hundreds of *funafuti* (lady finger) bananas, for which the atoll is named. Her menfolk carry bowl after bowl of seafood of every imaginable variety, including *'ota ika* (raw fish marinated in coconut milk), crayfish, crabs, tuna, and flying fish. The groom's table, meanwhile, is overflowing with the contributions from his family, including an endless variety of chicken dishes, loaves of fresh bread, and *pulaka,* cassava, and breadfruit dishes in all their pudding and porridge guises.

Emily works at the groom's table, and enlists my help removing the aluminum foil covers from one after another of the roasted pigs. There are big pigs and little pigs and littler pigs and gigantic pigs. Cooked whole, and just now brought out of the underground *'umu* ovens, they emerge from their silver wrappings perfectly intact, seemingly alive, except for the dehydration and the rictus smiles.

The platforms fill and overflow with food. Yet I am so mired in my chore, unpeeling pigs with the tweezers of my fingertips, that I don't immediately notice how this astonishing quantity of fare is only the tip of the iceberg. Lining the whole length of the *falekaupule* is a buffet line of collapsible tables representing the two families. The groom's people command the tables at one end of the hall, the bride's people command the tables at the other end, and, like restaurants, each offers specialties: seafood here, meat there, *pulaka* and breadfruit this way, fruit down there. These tables are likewise groaning under the weight of their provisions, including more roast pork — evidence that yesterday was a bad day for pigs on the atoll, as well as for their many weeping owners.

The bride and groom arrive with their bridesmaids and groomsmen. The Western clothes are replaced now with Polynesian finery: fresh gardenias and hibiscus and grasses interwoven with magenta-colored crêpe paper streamers and bright green-and-yellow fabric. The effect is dazzling, and the guests, who have been eagerly awaiting their arrival, pause in their food-preparation tasks to lean forward, greyhounds at the kitchen gates.

No one will eat until the bride and groom have tasted their own food, in fact, until the bride and groom have eaten a fair amount of their own food, followed a short while later by the honored guests at their tables eating a fair amount of the food. The general wedding guests, meanwhile, on the floor in perfect seated yoga poses, observe expectantly. Only the old men, leaning against the pillars supporting the *falekaupule*'s roof, appear nonchalant.

Fifteen minutes pass, and at a signal I can't perceive, the feast is unleashed as the tables along the wall of the *falekaupule* are suddenly obscured by scores of people using the tools of their hands to cut, saw, rip, pluck, twist, and tear at the whole pigs and the whole fish and the whole octopus and every other manner of food from the sea and from the sand. Both the restraint and the eating utensils shown at the first reception have been replaced here by a joyful, messy enthusiasm, and that which Rolf feared commences.

26
Just Do It

Y EARS AGO, WHILE working on a film about coral reefs, I covered a fete in Monaco sponsored by a glossy yachting magazine, designed to lure the world's super-yachts and their wealthy owners to a fundraiser for coral reef research and conservation on behalf of the Institute Oceanographie de Monaco. Prince Rainier had pledged attendance, and the Bal de la Mer was scheduled during the Fourth of July weekend, a holiday the Monégasque observe in honor of their very own American, the late Princess Grace.

One after another the super-yachts arrived at the Port of Monaco, a pocket-sized harbor nestled behind a seawall and visible from every street in the tiny principality-on-a-hill. Wooden boats, fiberglass boats, steel boats, some nearly a hundred years old, others barely out of dry dock, all in the 150- to 250-foot range, muscled a fleet of luxury (but not luxurious enough) yachts out of the berths. Looking down on the harbor, it seemed as if a herd of oceangoing mastodons had camped there.

The events of the ball included cook-offs among the yachts' famous chefs, contests for best interior decorating, and just about every other rivalry that might arouse the competitive juices of the rich and their impeccably uniformed crews. The contests unfolded within a bubble of opulence and *egalité* carefully orchestrated to ensure that all the yachts excelled at one thing or another.

That is, until a yacht of such enormous size that it could not fit inside the Port of Monaco appeared. Moored alongside the breakwater, *Lady Moura* arrived in the dark of night and greeted all of Monaco like a vision one morning: three hundred forty-five feet of steel as sleek and dynamic as one of Poseidon's stallions. The other yachts might as well have been digitally removed from the picture, since every human eye was drawn to this five-story cruise ship, whose gleaming white hull mysteriously opened and delivered forth astonishing things on hydraulic platforms: a red Ferrari, a forty-foot "cigarette" racing boat, dining terraces, sun decks. Several times a day all of Monaco was deafened by the sound of *Lady Moura*'s eight-seater helicopter roaring to flight to whisk the owner somewhere fun, down to Portofino for lunch, for instance.

This extraordinary boat belonged to the Saudi Arabian finance minister, although it was actually inhabited by his Lebanese wife, who was said to despise the restrictions of Saudi Arabia's *shari'a* law, and who refused to live there. *Lady Moura* was her home, a one-hundred-million-dollar floating palace complete with an onboard beach resort with a man-made sandy beach, an Olympic-size pool with retractable roof, a seventy-five-foot dining table bigger than most of the luxury yachts in the harbor (designed by Britain's eleventh in line to the throne, Viscount Linley), and the world's largest saltwater aquarium, complete with a living coral reef — the largest successfully nurtured in captivity at that time.

The Saudi Arabian princess was a well-known fixture in Monaco and Monte Carlo, and her arrival charged the air with an electrical expectancy: she was richer than rich and anything might come of being near her. All the crew on all the yachts pressed their navy-blue shorts and their white polo shirts a little stiffer, and all the shopkeepers of Monaco employed window washers so their goods would appear behind sparkling glass. The princess was preceded through life by brooms and bleach, making her view of the world different than yours or mine. Her charitable contribution to the Bal de la Mer was eagerly anticipated.

The last night of the fete was the night of the actual ball, and all the wealthy yacht owners, and all the wealthy denizens of Monaco's lenient tax laws, attended in formal wear, complete with featherlight shoes and gemstones the size of meteorites. The princess herself wore a gigantic emerald pendant that at first glance looked like a pet bat perched on her bosom. This was accompanied by an elegant floor-length gown and a black leather motorcycle jacket embroidered with an American flag, which had been prominently displayed in the sparkling window of one of Monaco's premier boutiques. The guests drank champagne and cocktails on the terrace overlooking the harbor, and overlooking their amazing boats made invisible in comparison to *Lady Moura.*

For the most part, despite the glamour, the prospective donors seemed bored. They were wearing their best but, in the end, what do you do in your best? They sipped drinks and smoked expensive cigarettes and sighed at the familiar ennui. Only the very young English wife of a much older wealthy Monégasque seemed in the least bit animated. Don't you think I should throw a fundraising party? she asked. What do you think would be better, something for a children's cause or something for an animal cause? Like the gowns she had donned and discarded in preparation for this evening, her charities were disposable.

To everyone's disappointment, Prince Rainier did not attend, although his son and heir to the throne, Prince Albert, did. Like many stutterers who struggle to complete a sentence in public, Prince Albert was liberated from his curse when he sang. And so rather than deliver the boilerplate royal speech at the microphone, he jumped onstage to join the band, belting out a passable rendition of James Brown's "I Feel Good." The crowd relaxed, and for a moment the languor lifted as everyone genuinely enjoyed themselves. This would have been the time to hit them up for donations. But apparently such things are not done among the superrich. And so the moment passed, to be followed by door prize drawings, an event that curdled whatever goodwill the prince had awakened.

Three prizes were offered, the winners to be randomly chosen from numbers on the name cards on the dining tables. The third and least impressive prize went to the owner of the third largest yacht. The second prize went to the owner of the second largest yacht. The first prize, a pair of round-the-world, first-class tickets on Delta Airlines, went to the Saudi Arabian princess. The crowd grumbled and redrew the cloak of boredom around themselves. Only the winners were happy. Genuinely and surprisingly happy, and I realized that even for those who have everything, the appearance of good luck, however rigged, is welcomed with child-like enthusiasm. Gosh, gushed the princess at the microphone, I've never flown commercial. I can't wait to try it.

After the Fourth of July fireworks over the harbor fizzled, after all the boats had slipped off for the next port of call, and after all the bills for the fete came due, the Bal de le Mer raised less money than the cost of the princess's leather jacket. The rich had not even emptied their pockets of their lunch money, and none of us in the film crew were surprised when word came round a year or so later that the Institute Oceanographie de Monaco had yet to spend the largesse. Among ourselves, we had come to refer to the event as the Bal de la Merde.

Now, years later, on the other side of the globe, the wedding feast in Funafuti is overflowing with pork, which everyone shares with greasy fingers and washes down with fresh coconut water fresh from the shell. We are entertained by the Pussy Cat Dancers, a troop of mostly *fa'afafine*, or transvestites, performing the more provocative Tuvaluan dances traditionally reserved for women. Clothed in coconut-shell bras and low-slung *lavalava*, hips shimmying, arms waving, casting seductive smiles, the *fa'afafine* draw cheers and whoops and squeals of delight from the audience. The wedding guests — men, women, and children — scuttle to the center of the *falekaupule* to reward a favorite dancer with a surreptitious spritz of perfume, or with money slipped under a bra strap. The groom and groomsmen award the dancers for every song, and the crowd shouts enthusiastically each time.

Ironically, in repressing Polynesian dance for women, the Christian Church of Tuvalu has unwittingly cultivated its resurgence among transvestites.

After the Pussy Cats, and in between the business of eating, the guests themselves take to the floor, old and young, fit and feeble, dancing a hybrid hula you might see at a club in Honolulu, only slower. The difference is that here in Funafuti there is no physical contact whatsoever, and partners refrain from even looking at each other as they shimmy sedately from one hip to the other, arms jogging lightly by their sides. The young women particularly have developed a dance so subtle as to be almost subliminal. Only the older women cut loose, loudly asking young men to dance, particularly the groom and all his groomsmen. Emily proves the wildest dancer of all, throwing her arms and legs around like a fighting hermit crab. She is rivaled only by an ancient woman whom I've kept an eye on all day, worried she might not survive the festivities. Somnolent and almost comatose, this prehistoric *fafine* (old woman) snaps to life now, forging onto the dance floor, where she proves to be not a dynamo, exactly, but a shining embodiment of her Nike T-shirt that says *Just Do It*.

Underlying all the fun and the food and the music is the knowledge that in two days' time the newly married couple will return to their university studies in Fiji. The bride's father rises to give a speech, tears in his eyes. She is his only daughter among five sons, Emily tells me. The father speaks at length, unabashedly emotional. It doesn't matter that I can't understand Tuvaluan, I know what he's saying: his daughter is leaving his household. She is doing what all Tuvaluans are thinking about doing whether they admit as much to *paalagi* visitors or not. She is stepping into a future that will likely carry her to a new home, a reality the Tuvaluans have been spared for the past two hundred centuries.

Meantime, the bride and her new husband tear into the pig carcasses with their bare hands. They are two of a growing number of Tuvaluans who do not live in Tuvalu, and are a tiny embodiment of that which is reflected in the most recent census, showing the population down nearly 25 percent from a high of more than twelve thousand. The other thousands have already fled the rising waters, or the limited opportunities, and are now scattered across the South Pacific, many in Auckland, the largest Polynesian city in the world, and a place decidedly pigless and landless, at least for refugees.

27

Beseeching the
Wind Horses

I N DESCRIBING HIS PEOPLE to the world, former
prime minister Faimalaga Luka wrote, "Tuvaluans are
blessed, as divine right or good fortune would have it.
We have the sea, and above all we have our land. [We] are closely
knit through kinship, a small population, and a single binding culture. What this mixture stirs up is a sensation that runs deep, a
supreme sense of place."

Since its independences in 1978, Tuvalu has behaved differently
from its South Pacific neighbors, many of which are among the
most corrupt nations on earth. But democratic Tuvalu has managed its resources well. It has grown the national trust fund
bequeathed to it by Britain, Australia, and New Zealand upon independence from twenty-one million dollars to more than sixty-
three million. It has sold commercial tuna fishing licenses within
its Exclusive Economic Zone. It has produced postage stamps for
the international philatelic trade and licensed its Internet country code (.tv) for $12.5 million, using these monies to purchase
Tuvalu's first streetlights, pave its one and only road, and buy
membership in the United Nations.

Other schemes, though lucrative, are more controversial, such
as selling Tuvaluan flags of convenience to shipping interests (including a Tuvaluan-flagged North Korean freighter impounded
by Australia and found to be carrying fifty million dollars' worth

of fundraising heroin reputedly for Kim Jong Il's nuclear weapons program). Tuvalu also once sold passports, allowing anyone to become a citizen for a mere eleven thousand dollars, a scheme that died after terrorists became Tuvaluans. For a few lucrative years Tuvalu even sold its 688 telephone country code to a telephone sex service, a business that once earned 10 percent of the nation's federal budget, yet was forsaken in the wake of objections from the nation's Christian churches.

The country's biggest potential revenue earner, however, lies in a lawsuit against the United States and Australia to be heard by the International Court of Justice in The Hague. Tuvalu is recruiting other island nations to sue for damages from climate change, arguing that the small island nations of the world, which contribute only 0.6 percent to global warming pollution, disproportionately suffer its effects. In the course of international jurisprudence, Tuvalu hopes to penalize the rich world for practices that essentially amount to loosing their pigs in their neighbors' gardens.

Beyond Tuvalu's shores is skepticism over this lawsuit, with some contending that it's nothing more than a ploy for foreign money. "We hope only to speak for the low-lying atolls and coastal areas of the world," says Prime Minister Sopoanga, although he admits that the lawsuit is facing an uphill battle from litigants who fear reprisals from the powerful donor nations they would be suing. Nevertheless, the Tuvaluan case is considered threatening, and Australian legal experts, at least, have advised their government to take it seriously.

Along with Tuvalu, many other island and coastal cultures have just grievances about sea levels. Kiribati, Tuvalu's neighbor and former partner in the Gilbert and Ellice Islands, has already lost two *motu* to the rising waters. The seas around the Carteret Atolls off Papua New Guinea have cut one island in half, leaving fifteen hundred people at least temporarily dependent on food aid. Gravesites in the Marshall Islands, including World War II graves, are washing away. The island nation of Trinidad is losing shoreline at the rate of seven feet a year. A third of the two hundred inhabited atolls of the Maldives in the Indian Ocean are in danger of disappearing. Throughout coastal Bangladesh, villagers are attempting to raise the height of their islands and shorelines with buckets of mud dredged by hand from offshore. Even far from the tropics, on solid continental land, people are feeling the heat, as Inupiat Eskimo villages slip from the rapidly melting tundra into the sea.

Help has been promised, but it pales in comparison to ongoing practices. In 2001, the rich nations of the world pledged $0.4 billion a year to help developing countries adapt to climate change, though the offer was overwhelmed by the eighty billion dollars the wealthy nations spend annually on energy subsidies for fossil fuels. In the Pacific, the frustration is apparent. "Tuvalu's voice in the debate is small, rarely heard, and heeded not at all," writes former prime minister Faimalaga Luka.

As a visitor to Tuvalu, you might find yourself wondering what Charles Darwin would have thought of it all. In the course of his long travels through the Pacific he gleaned much about evolution and its shadow-partner extinction: "We have every reason to believe . . . that species and groups of species gradually disappear, one after another, first from one spot, then from another, and finally from the world." Perhaps in light of today's realities he would construe the same fate for vulnerable human cultures, or even for all of human existence.

In the most ancient roots of Indian thought, predating even the 4,500-year-old (or older) scriptures known as the *Vedas* (from the Sanskrit *ved*, cognate of the English words *vision* and *wisdom*), exist spirits known as *nagas,* variously described in Hindu, Jain, and Buddhist mythology as cobras, dragons, or serpents. Sometimes the *nagas* are half-man, half-serpent, and sometimes they live in the fantastical underground city of Bhogavati. But often the *nagas* and *naginis* (female) inhabit the world of ponds, lakes, rivers, streams, and oceans.

For Tibetan Buddhists, the *nagas* are landowners who possessively guard their homelands and homewaters from the encroachment of humans. When people do intrude — draining the waters, or polluting the waters, or digging construction pits, or plowing fields — the weak *nagas* grow frailer, perhaps dying off into extinction. But the strong *nagas* wax powerful in their anger and vengefully invoke epidemics or natural catastrophes. They rarely assail the true offenders, however, whom they can't identify, instead besetting any person or people already down on their luck.

Over the centuries, the people of Tibet have negotiated all kinds of ways to soothe these savage spirits. As a general prophylaxis, they fly prayer flags bearing images of horses to keep their wind-horse running, namely, to improve their luck. Far away in Tuvalu, the people are bereft of wind-horses, although their *fale* are generally graced with the fluttering colors of freshly laundered *lavalava* drying in the wind. Perhaps these streamers will lure luck their way, while redirecting the ire of the *nagas* elsewhere.

An old man sitting next to me on the floor at the wedding reception is wearing a *lavalava* stenciled with images of eels. When I ask him about it, he tells me about Tepuhi, the Tuvaluan sea serpent spirit. At the beginning of time, when the heavens and the earth were united as one, Tepuhi lifted the sky onto her back and pressed it into place overhead. Looking down, she saw the earth was now one big block of land, not at all to her liking. So she smashed it into little pieces, enabling the ocean to enter the space between the newborn islands.

This ancient *tagata* (old man) says that Tuvalu has always had good luck because of its eels. When I ask whether Tepuhi might have anything to do with the rising waters and the possible submersion of Tuvalu, he declares that he will never leave, and if need be he will take to his *vaka* and paddle it around, eating sunshine until the end of his days.

Five hours or so into the second wedding reception, I realize that the dancing and the feasting are only getting started, and the

Tuvaluans are pacing themselves on a time scale I cannot follow. Without doubt I am an amateur at feasting, and without doubt I am an amateur at sitting on the floor. Hobbling back to the Hideaway Guesthouse on sore legs, I pray that no one will drive past and offer me a ride I cannot politely refuse. For once, I would like to walk the six miles home.

Luckily, only Father Kamilo passes, in the van he uses to collect his flock for Sunday Mass. He is on his way to the wedding to pay his respects, and to offer a gift, even though the newlyweds and their families are not Catholic. But after more than sixteen years on Funafuti, the Father is part of the *fenua,* the neighborhood. He understands when I refuse a ride, but insists on getting out of the van and, laughing, shows me his shoes (sneakers), which he never wears. He wears only the flip-flops that he calls, variously, flop-flips or flap-flips. After years of practice, he can "flip without flapping" — a feat we agree is a notable achievement on the long road to tropical acclimatization.

The last mile or so I walk through the rain, enjoying the sight of Tuvaluans sitting outside, chatting and laughing as if the sun were shining. Even the pigs under their little roofs are leaning against the windward side of the bars trying to get as wet as possible. When I arrive at the lodge, I see that Rolf is also drenched, but he is looking sad and confused. He shows me how all three of the mother cat's little kittens have been dragged out from under the sofa in the lounge and killed by a dog. Their tiny bodies are scattered along the walkway outside the front door: one orange kitten, one black one, and one calico. It's the first we have seen of them, and he and I stare at the lumps of wet fur, while the thunder peals, and the church bells peal, and the fronds of the coconut palms hiss in the wind.

She has lost her little kittens, Rolf keeps saying. And because he is bereft, I collect the bodies and dig a hole with a clam shell in the sandy soil behind a mango tree and bury them there in the rain.

Emily remains at the wedding, eating and drinking and dancing, mostly without sleep, for the next two days and nights. Rolf grumbles that after thirty years of marriage she never sleeps at home anymore. Concerned, I keep him company and cook him a pot of fish soup with canned peas, which he pretends to sip before heading shakily to his bedroom, a stringer of beer in hand. This night, unlike most, the rain falls for hours, the sound on the corrugated metal roof drowning the thunder of surf. For the first time since my arrival, Foliki does not join me under the *lavalava*. Sleepless, I worry. But in the morning I see that her fondest wish has come true, and that her mother, dispossessed of her littlest kittens, has allowed the sole survivor of her previous litter to resume nursing.

28

Sinking Dragons

O F THE MANY SCIENTIFIC projects measuring atmospheric and climatic changes, one of the most comprehensive tracks eighteen hundred robotic floats seeded in the oceans and connected to satellites orbiting overhead. These robots, collectively known as Argo, rise and fall through the top 1.25 miles of the world ocean, continuously measuring variations in temperature, salinity, gas composition, and velocity.

Among those analyzing Argo's wealth of data are NASA climatologists, who have found evidence of a fundamental shift in the planet's energy exchange system, with the world ocean now absorbing — for every ten square feet of sea surface area — one watt more of the sun's energy than it radiates back to space. James Hansen, the report's lead author, refers to this 0.85-watt differential as the smoking gun of global climate change because it corresponds closely with predictions made by supercomputer climate models.

According to Argo, we are not riding into the evolutionary future on a swift and nimble *vaka*, but rather on the energy equivalent of an ocean freighter, burdened by thermal inertia so enormous it cannot be redirected in a time span of less than centuries. Two computer-modeled studies out of the National Center for Atmospheric Research in Boulder, Colorado, predict that global climate change is now unavoidable, and that sea level rise will outpace atmospheric temperature rise both in intensity and duration because the oceans are slow to absorb the heat in comparison to the atmosphere. Since the heat trapped in the oceans accounts for most of the energy the earth has absorbed from global warming in the past one hundred years, sea levels will rise, and possibly accelerate, for at least the next five hundred years — rendering a completely new map of the world, as river valleys become seas, continents fragment into islands, and thirteen of the world's twenty most populous cities submerge.

The modern Argo array is named for the oared ship of war (not unlike the Polynesian *vaka*) that Jason and the Argonauts sailed from the ancient Greek city of Iolkus in search of the Golden Fleece. Its name derives from the Greek word *argos*, which derives from the Sanskrit *rjrah,* meaning "shining, glowing, bright" and "swift." The Greek ship was no mere vessel. Aided by a remarkable "speaking timber" taken from the sacred oaks at the Oracle of Dodona and fitted into *Argo*'s bow by Athena herself, *Argo* offered all kinds of advice useful for keeping on course through storms, battles, and every manner of treachery the ancient world could muster. Yet the ship eventually proved Jason's undoing. Rotting from the neglect of years, the timbers of her mast fell, killing him and destroying her in one all too preventable Greek tragedy.

Science, at its best, is an oracle, although it too is often ignored.

Thoughts of *Argo* accompany me as I paddle across Funafuti's lagoon on an ancient *vaka* I dig out of the jungle in the Hideaway's back garden. Och, it's ruined, says Rolf as I drag it down to the beach. It's finished, he calls after me. But I am not so sure. I bring my snorkel gear in case the vessel sinks and I have to swim back. Rolf hands me half a coconut shell — for bailing, he laughs. Yet *Tepuhi*, as I call her, skims across the water, easily bypassing the impenetrable thickets of staghorn corals, the bone graveyards below.

Under the water / what is the grebe's heart like? asked Izumi Shikibu one thousand years ago. So it is between the underwater and the topside worlds: that which we know up here, we cannot know down there. Through the skin of the surface, the deeper water in Funafuti's outer lagoon is iridescent, the blue-black fading to midnight near the bottom. The water is so clear I can make out coral heads sixty feet below, which appear, from the *vaka,* as black islands clothed in thick forests, and trembling with birds (fish). When I glance back, Rolf is standing on the shore in the shade of a coconut palm, waving a long goodbye.

Did you see the boat? he asks when I return.

The boat?

The boat we used to drive out to the *motu,* he says.

His memories resemble plankters adrift on a gray neural sea, arising on this wave crest, dipping into that trough. They come and go, and sometimes the visibility is good and they appear as sharp as an albatross on stiff winds, and at other times they are as elusive as sea spray. An hour ago Rolf could not remember ever having a boat, not even the *vaka* I've just paddled back to shore.

A boat you used to take to Tepuka Savalivili? I ask.

But he only looks confused, as if he's never heard of the disappearing *motu.* Och, no, he isn't sure which *motu.* A pretty *motu.* One with coconut palms and coconut crabs. They had a boat and it sank out there during a storm, he points. No, he corrects himself, it sank when another boat hit it. He can't remember. It was called *Akiaki,* for the fairy terns that once nested on the unfinished roof beams of the Hideaway Guesthouse.

On my last night in Funafuti I offer to take Emily and Rolf out to dinner in the village. Emily is eager to visit the one and only Chinese restaurant in the country, but whispers to me that we should leave Rolf at home because he will be a "headache" to us. Rolf, who is getting weaker daily, and who eats nothing anymore as far as I can tell, will not hear of it and insists on joining us. We head off to our respective rooms to dress, and when I meet Rolf again in the lounge, he has donned white cotton slacks and boat shoes and brushed his hair. Are you coming to this function, too? he asks.

This function? I ask.

I don't know what it is, he says. Is it a party?

It's dinner, I say. I'm taking you and Emily to dinner.

Unruffled, he leads the way roadside and flags a passing car.

The Vailiki Chinese restaurant is run by Tuvaluan friends of Emily's on the upstairs deck of their house, and Rolf cannot climb the exterior wooden stairway. Och, he scoffs, I can get up there. But his legs won't work. Emily, determined, protective, exasperated, loving, asks me to get ahead of him on the stairs and lift his feet, one at a time, and place them on the steps. Standing behind him, she takes his weight against her body and, holding both Rolf and herself in place with the strength of her arms clamped to the handrails, levers him upward with her hips. It's a slow and painful means of ascent, and Rolf complains the whole way: This is stupid, for Christ sakes, I can do it, after thirty years of marriage, she won't even let me walk on my own two feet anymore. But he is helpless to help himself or to fight us.

We wait a long time for our beef chow mein and our fish curry and Rolf's egg drop soup to arrive. Rolf tells us how much he likes egg drop soup, and how much he doesn't like everything else on the menu, which is *so much food* and *so much fish*. But when dinner arrives he just stares at the egg flowers swirling and congealing in a clear broth, his body too wracked with hiccups to eat. For once, he falls silent . . . then rallies a half-hour later to tell a tale of how Emily's family came over to dine and one of her aunts ate ten flying fish.

Standing on the threshold of the Hideaway Guesthouse, Rolf places a Tuvaluan necklace of cowry shells around my neck and says he hopes I will remember this place. Emily hugs me. Shortly afterward, rolling toward takeoff, the plane passes clusters of small children lining the runway, their mothers standing beside them, infants on their hips, everyone waving.

The rickety twin-engine turboprop is filled with patients destined for hospitals in Fiji and beyond. The daunting work of maneuvering these ill Tuvaluans on their stretchers, and in their wheelchairs, up a tiny gangway and down the narrow aisle has considerably delayed our flight. Yet, no matter what their medical conditions, they have been delivered to their assigned seats and made fast with seatbelt extenders. Most are only semiconscious, a few look close to death, and, as far as I can tell, I am the only one aboard waving back at the children.

When we bump into the air, in no time at all, the crowded, dirty, changing, hopeful, fearful realities of Funafuti fade into the image of a white dragon swimming against a sapphire sea, its sinuous body festooned with green coconut-palm feathers, its back embellished with the racing stripe of a new road. Crossing to the far side of the atoll, all the tiny, uninhabited *motu* become the knuckles of the dragon's tail, partially submerged, fighting the windward combers. Then we're over the open Pacific, where the tradewinds are ripping whitecaps a quarter mile long, and I feel the arrival of *yugen*. The patients aboard are not looking back, and who knows if they will ever see Tuvalu again.

Within the coming decades, many, if not most, of the earth's three hundred thirty atolls will likely revert to sandbars, and then to nothing. Although the people themselves will not go extinct, without their home islands to anchor them, their beliefs and their identity will. Scattered person by person across the rising waters, they will migrate to places where they will learn to wear real shoes and to eat frozen pork until, like Atlantis, their story fades into myth.

Part III

Mo'orea
Society Islands

In the ocean are many bright strands
And many dark strands like veins that are seen
When a wing is lifted up.

— *Mawlana Jalal ad-Din Muhammad Rumi*

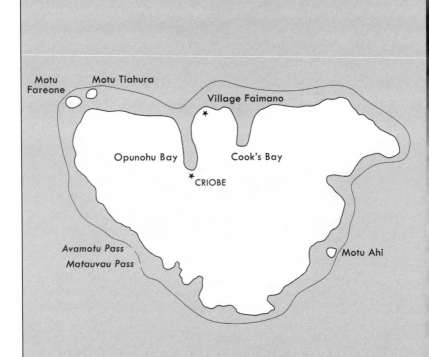

Motu
Fareone

Motu Tiahura

Village Faimano

Opunohu Bay

Cook's Bay

CRIOBE

Avamotu Pass

Matauvau Pass

Motu Ahi

0   3
Approximate Miles

MO'OREA

29

The Churning
of the Ocean

S HE IS BOBBING through the calm waters inside the
barrier reef, near the swift channel between two *hoa,*
where a pair of streams enter the sea. I am loitering
in the calmer water myself, dipping and fluttering my canoe paddle and trying to steer a straight course. It's just after dawn. The
air has not yet awoken to the convection currents that will drive it
later this day, and the lagoon is motionless, almost oily looking,
with a glycerin sheen as absorbent to color as a paintbrush. In the
ripples to my left, the peridot green of Mount Rotui blooms. In
the ripples to my right, the obsidian blue of the open sea. Between the two lies a surface as good as a window, through which
sergeantfish, anthias, and lionfish drift above little islands of corals and gatherings of stingrays.

From afar, she looks like one of those ubiquitous pieces of
oceangoing flotsam washed from shore or ship and plying the
ocean with indestructible endurance. I paddle toward her, bent
on litter collection, only to discover that she is not a styrofoam
cup or a plastic sandal but a living creature roaming inside her
own home — an argonaut, or paper nautilus, probably of the species *Argonauta argo.* She is a member of a genus of octopus that
long ago abandoned life on the sea floor in favor of roaming the
open ocean. Unlike her namesake, the chambered nautilus, her
delicately coiled shell is not an external skeleton that she is at-

tached to as we are to our fingernails, but a mobile home that she can come and go from like a hermit crab.

I have never seen an argonaut alive in the sea before, and with fumbling hands I don mask, snorkel, and fins and slip over the side, dragging the *va'a* by the float so as not to lose it. She is a timid creature and this may be the only opportunity that ever comes my way to see her in the wild. I waft my fins gently, knowing that I must approach as softly as a ripple.

It doesn't matter. She is engaged in one of those acts of violence that nearly preclude thoughts of personal safety. She is half out of her shell, pulsing in bright red and yellow, the colors literally tumbling through her like reflections from flashing police lights. Her colors are so strong that they bleed beneath the skin of her paper-thin shell, bruising it. She is administering the coup de grâce to a pteropod, a sea butterfly. Her eight arms are flared open, an umbrella turned inside-out, exposing the parrotlike beak. The pteropod is flapping its transparent wings in hopes of escape but the argonaut is reeling it in on the sucker disks of her arms, biting it, then tucking it under her bell, and rolling herself back into her translucent shell, where the flames of her hunting colors soften to pink.

Quietly now, her big eyes innocently wide, she floats a foot below the surface, arms wrapped over her head, the tips of them tucked daintily into her shell, leaving most of her sucker discs exposed. She observes me from a safe distance, one orange eye watching me as she feints toward shore, the other watching outward as she tacks toward home in the open sea.

As with every female argonaut octopus, she has fabricated her shell herself, using calcium carbonate secreted by the two large, flattened dorsal tentacles unique to her genus and gender. She has invited a much smaller — no more than an inch long (invisible to me) — father-to-be aboard. Now, with her head and tentacles protruding, she sails herself, her mate, and eventually her brood around the tropical and semitropical oceans of the world. Carrying the family on the currents, her shell is as fragile as a parasol of bone china, yet strong enough to shield its occupants from the ultraviolet radiation present near the surface of the oceans.

In his *Historia Animalium,* Aristotle described what was known, twenty-six hundred years ago, of the life of the little argonaut, noting peculiar talents associated with the females' specialized dorsal tentacles:

> It rises up from deep water and swims on the surface; it rises with its shell down-turned in order that it may rise the more easily and swim with it empty, but after reaching the surface it shifts the position of the shell. In between its feelers [tentacles] it has a certain amount of web growth, resembling the substance between the toes of web-footed birds; only that with these latter the substance is thick, while with the nautilus it is thin and like a spider's web. It uses this structure, when a breeze is blowing, for a sail, and lets down some of its feelers alongside as rudder-oars. If it be frightened it fills its shell with water and sinks.

My *fare* at the Village Faimano on the island of Moʻorea contains a little fleet of three paper nautilus shells decorating the windowsill above the kitchen sink. I imagine this translucent flotilla stranded on storm waves or on rogue winds, and that the hapless octopus families died on one of Moʻorea's infant *motu.* In another time and place these shells would have dissolved to sand, and, with the help of seabird guano, eventually transformed to beach cement and a part of the growing *motu.* But for the time being they are living out their afterlife in the kitchenette of the *fare,* alongside all the other beach treasures that will someday return to the sea via the same dispassionate vehicle that stranded them.

This island of Moʻorea, where the argonauts came to land, is one of the Pacific Ocean's pelagic traps, luring those who need landfall and those who don't, and delivering them to either salvation or doom. The island itself is one of the fourteen or fifteen (the number varies with the tides) islands that make up the Society Islands of French Polynesia, including the well-known islands of Tahiti and Bora Bora. Mo'orea is one of the five islands the French call Îsles du Vent, the Windward Isles, because they are situated ahead of the Îsles Sous le Vent, the Leeward Isles, on the path of the southeasterly tradewinds.

Mo'orea lies four hundred miles southwest of Rangiroa and twenty-two hundred miles southeast of Tuvalu. The island is all that remains of a once fully active shield volcano 1.2 million years old, which once rose ten thousand feet into the tradewinds, where its summits sheared enough moisture from the fleecy clouds to sustain rushing rivers. What survives today is the volcano's southern rim, a dramatic fifty-square-mile triangle of land intensely eroded into spires and sawtooth ridges nearly four thousand feet high. Assuming its future continues to include coral reefs, Mo'orea will someday become a coral atoll, like Rangiroa and Funafuti. If not, it will become a flat-topped guyot, lost beneath the waves.

The topography of this small island is among the most dramatic per square foot of any on earth. Mo'orea's northern face is cloven by two long fjordlike bays nearly enclosed by rugged peaks rising sheer from the sea. Opunohu Bay in the west is flanked by the mountains of Tautuapae, Mouaroa, and Rotui, and Cook's Bay in the east by Rotui, Mouaputa, and Tearai — each rearing between twenty-five hundred and three thousand feet above sea level. From certain angles, Mount Mouaputa winks an eyelike hole that perforates its summit, said to have been pierced by the spear of the demigod Pai.

Mo'orea's remnant volcano is currently wrapped in green jungles decorated with hibiscus, *tiare, hutu,* and *i'ita* flowers, along with countless species more recently imported on waves or boats. The entire island is surrounded by a cocoon of surf that breaks on the barrier reef. These combers underline in white the ancient shoreline of Mo'orea's volcano, while the lagoon highlights in aquamarine the mountainous flanks that have already subsided beneath the waves. At this point in its transformation, Mo'orea has accrued three small *motu* atop its barrier reef — Fareone and

Tiahura off the northwest tip, and tiny Ahi off the east coast. These sand islets offer a glimpse of what might someday surround the lagoon of the future atoll.

Unlike Rangiroa's barrier reef, which is broken by two navigable *hoa*, Mo'orea's barrier reef contains twelve passes, all of them navigable by small craft in fine weather, and one big enough to admit large ferries from the neighboring island of Tahiti, ten miles to the east. Each of Mo'orea's *hoa* corresponds to a freshwater stream that runs from the mountains, or once did, and each of these *hoa* admits many open-ocean travelers into Mo'orea's shallow lagoon, including the wayward argonaut.

Dragging my canoe in the water behind me, I accompany the paper nautilus on her dangerous journey inside the barrier reef, imagining that I will shepherd her to safety in open water if she looks in danger of stranding. In languid slow motion she ascends and descends within three feet of the surface, her colors discharged to a serene and nearly translucent white, her shell as lustrous as a freshwater pearl. She appears to be asleep, tucked inside her mobile home. She is no bigger than my open hand, yet she has attracted a tiny crowd of shade-lovers who flutter below her: a school of larval fish too small for me to identify.

Here, on the leeward side of the island, there is no surf breaking on the barrier reef, only humps of water flexing their muscles over the berm of the algal ridge. Our tempo quickens as we are drawn on an invisible back current running through Teavarua Pass, where the swells of the pelagic ocean heave and relax in breathlike rhythm. The argonaut bobs with this new cadence, her coterie of shadow-dwellers keeping perfect time below.

Backlit against the slanted rays of morning light, she is surrounded by hordes of ctenophores — jellyfishlike invertebrates the size of grapes and flashing rainbows of light through their cilia (Latin *cilium*: eyelash). Tucked into the swirl of her shell, the argonaut floats faster than her fellow drifters, passing zooplankton and transparent egg cases that look like cellophane noodles. Bobbing into a tangle of these, she plummets to a depth of twenty feet. A moment later I understand why. Strands of these egg casings wrap around my lips, cheeks, neck, and ears, imparting mild stings. When I try to unwrap them, they sting my hands and cling there, too. Unnerved, I haul myself back aboard my *va'a* and wipe a towel across my skin, then dip the towel into the sea, hoping to set these embryonic travelers back aboard their current.

The last I see of the argonaut she is floating ten feet underwater, sailing her lustrous vessel on watery winds, headed toward the westerly islands of Samoa.

The *Mahabharata* of India, one of the two longest poems on earth, with one hundred ten thousand couplets, recounts a secondary creation myth known as *Samudramanthana,* or the Churning of the Ocean. Gods and demons, weakened from one of their fourteen efforts to recreate the universe, decide to mine the ocean — the vault of all potentialities — for *amrita,* the nectar of immortality.

Together these cocreators uproot Mount Mandara to use as a stirrer. Vasuki, the king of the *nagas* (serpent-spirits), volunteers to coil himself around the mountain. He allows the gods to pull his tail, the demons to pull his head, and, as if rubbing a firestick between two palms, they stir the ocean. For ages the gods, the demons, the mountain, and the *naga* churn until Mandara bores through the ocean floor toward the center of the earth — an unforeseen development that threatens to undermine their efforts. Yet all is not lost. Coming to the rescue, Vishnu the Preserver, assuming his savior *avatara* (incarnation) as Kurma the Tortoise, dives to the bottom of the sea and offers his shell as a platform to bear the weight of the churning Mount Mandara.

Thus, one by one, the treasures emerge from the abyss, including Surabhi, the cow of plenty, Varuni, the goddess of wine, and Lakshmi, the goddess of fortune and beauty. Chandra, the moon, comes up, as does Uchchaisravas, the seven-headed white horse, Airavata, the three-headed white elephant, and Sankha, a conch shell. *Amrita,* the nectar of immortality, is carried to the world by

Dhanvantari, physician to the gods, and is followed by the *apsaras* (divine beauties and celestial dancers), who appear with their favorite tree, the *parijata,* or coral tree, whose dramatic red flowers bloom when the tree is leafless, perfuming the world. Thus all the potentials are manifest.

On the grounds of the Village Faimano, just outside my *fare,* stands one of these coral trees, which the Tahitians call 'atae, and the French call *arbre immortel,* whose orange-red flowers explode like clusters of flames amid bare branches. At dawn and dusk these flowers release an incenselike fragrance that attracts the yellow birds known as wattled honeyeaters. Noisy, aggressive, and quarrelsome, their antics in the 'atae resemble little gods and demons fighting over the birth of each dawn and each dusk, and in the wake of their theatrics the sand beneath the tree is littered with shredded red flowers.

This 'atae originates from far away. Long ago the seeds of its ancestors, perhaps caught in the churning of a great typhoon, washed from a distant Asian shore and crossed the sea, surviving their time afloat thanks to their canoelike legumes: the six-inch-long seed pods carrying half a dozen egg-shaped seeds as glossy and red as beads. At least one hundred ten coral tree species share the genus *Erythrina,* and this specimen outside my *fare,* a member of *Erythrina variegata,* is one of the most widespread, thanks to seeds that can survive a year or more in the salt water of the oceans.

Mo'orea's barrier reef strains some voyagers from the ocean while allowing others to continue on their way. The 'atae seeds that bumped ashore, and doubtless still bump ashore to colonize this island, eventually grew into tall trees. They and their descendants flourished where the soil was good, providing nectar for honeyeaters blown in on typhoon winds. When the first people arrived here sometime around the year 800, they discovered that the lightweight, buoyant wood of the 'atae — so perfect for making the floats on their outrigger canoes — had already colonized ahead of them.

30

An Ocean of
Silence and Bliss

T HE LITTLE *VA'A* I paddle in Mo'orea's lagoon is a modern construction made entirely of fiberglass, including its float. Yet its design retains the lightness and efficiency of the *'atae* wood, so that even my inexpert paddling allows me to skim with ease across large expanses of the lagoon in search of treasures stirred up by the churning of the oceans. In particular, I am hoping to find those travelers who cross through Mo'orea's barrier reef after dawn to spend the day resting in the lagoon.

Deep in Opunohu Bay, I spot their telltale ripples. They are so quiet as to almost be invisible: twenty-five or so spinner dolphins (*Stenella longirostris*) rising in tandem to the surface, the gray domes of their heads barely parting the surface. Each breathes once or twice without any noticeable spout, then sinks to the bottom, where the entire group swims in silence for a few minutes before rising steeply to breathe again.

They surface behind me, and on the next breath, they surface far to my left. In fact, where they emerge is so unpredictable, and their signature at the surface so minimal, that on several occasions I miss them entirely. Paddling my canoe to the left and to the right, I search every horizon, yet spend long minutes without results. Motorboats come and go. Up near the pass Jet Skis race back and forth. A pair of Picasso triggerfish — outlandishly abstract in white, brown, blue, and yellow — inquire through the surface with hungry eyes, hoping for scraps of bait. But I am fishing with only my senses and have nothing to give.

The next time the spinners breathe, they angle up from directly below my canoe, so that I can see through the water their tight formation, the dolphins swimming flipper-to-flipper in twos, threes, or fours, the whole rising as a long column of dolphins. One after another the subgroups surface twenty feet to my left. Yet the most I hear of their breathing is a slight aspiration, as if I put my lips together and quietly sounded the letter *P*.

From this close, their gray, white, and black markings stand out sharply against the white sandy bottom. This might seem counterintuitive to a safe sleeping strategy, but just as with the gray reef sharks schooling in Tiputa Pass on Rangiroa Atoll, the spinners have carefully chosen the parameters of their daytime bed. Foremost in their consideration is the presence of this sandy bottom, the white sheets of the sea, as crisp and clean as linen.

For dolphins of the genus *Stenella*, deep sleep bears little resemblance to what we do when we lay ourselves down and succumb to a state of autonomic paralysis, hypnagogic hallucinations, and dreams. Tucked behind the walls of Moʻorea's barrier reef, moving slowly en masse over a sandy bottom, the dolphins are alert and mobile, only in a different way than during wakefulness. Asleep, they are resting their hearing brains, while relying on their seeing brains to monitor the world around them — the reverse of what we do when we close our eyes to sleep and leave our ears open.

For these spinner dolphins, sleeping against a white sandy bottom is something akin to sleeping with the lights on, since it allows them to visually detect any predator (shark or possibly killer whales) coming their way. Traveling back and forth through Opunohu Bay, this school is carefully avoiding the mottled areas of the lagoon floor composed of fringing reef. They are also avoiding the head of Opunohu Bay, which is muddied by runoff from pineapple plantations in the Opunohu Valley. Given a choice, these sleeping spinners choose clear water and a sandy bottom in shallow water behind the fortification of the barrier reef. And this is where they come every day just after dawn and leave every day before dark — if not this bay, then another bay meeting the same criteria on a different side of Mo'orea's triangular landmass.

Studies of great white sharks hunting seal rookeries in northern California reveal that one of the sharks' primary hunting strategies involves careful utilization of different substrates in the sea. When not hunting, white sharks cruise over a sandy bottom, allowing the seals to see them and habituate to the sight of them. But prior to an attack, a shark camouflages itself against a rocky bottom, where its coloring makes it nearly invisible. It awaits the appearance of a seal swimming overhead then launches upward at great speed, and at a steep angle. The ambush, if successful, culminates in a single bite, after which the shark retreats and waits for its prey to bleed to death.

Although there are few great white sharks in the waters off Mo'orea, their tropical equivalents abound, the tiger sharks and great hammerhead sharks, both of which are capable of exceeding eighteen feet in length. The Tahitian spinner dolphins, only five-feet long and one-hundred-fifty pounds, are as vulnerable to these enormous hunters as you or I would be, although they have evolved ingenious means of outwitting them, even during sleep.

As far as is known, all dolphins sleep only one brain hemisphere at a time, allowing the other half to rest, a technique that enables them to continue consciously breathing at the surface. Since cetacean respiration is not autonomic, they must control every phase of it. Consequently, these spinners never lose consciousness during sleep, as we do. In fact, unconsciousness, as we know it, proves fatal to whales and dolphins. Hence their proclivity for rescuing their unconscious brethren by buoying them at the surface or by stranding — behaviors cetaceans sometimes employ for the benefit of faltering human swimmers.

The spinners asleep this morning in Opunohu Bay are tracking back and forth in a style scientists call the carpet formation, all the subgroups drawing close and maintaining the same depth just a few feet above the bottom. As they approach the edge where the sand gives way to fringing reef, with its potentially dangerous mottled bottom, they turn and reorient themselves over the sand. In this way the school crosses and recrosses the entire contiguous area of the sandy bottom, drawing invisible lines across it in much the same way a screen saver on a computer bounces inside the edges of the screen. The size of the sandy bed in any given bay will determine how many dolphins can sleep in it together.

But perhaps the most remarkable behavioral aspect of these sleepers is the silence of the schools. Anyone who has spent time in the water with spinners knows these are among the most talkative of cetaceans, and their presence in the sea transforms the near-field into a collage of richly colored and textured sound. At different stages of the day and night, depending on what they are doing, the spinners' world is a near-deafening collection of whistles, click-trains, and a strange variety of burst-pulsed signals that sound to our ears like a barnyard orchestra of quacks, moos, baahs, blatts, banjo twangs, barks, squawks, and chuckles.

Compared to other dolphins, including other members of the genus *Stenella,* the spinners have a wildly exuberant soundtrack that occurs well inside our own hearing range and sounds uncannily like a human language we vaguely recognize but don't understand. The late Ken Norris of the University of California at Santa Cruz, along with his colleagues and students who have studied Hawaiian spinners for more than thirty years, dubbed one of the spinners' noisiest phonation displays the Yugoslavian News Broadcast.

But here in Opunohu Bay at midday, the *petite* Tahitian spinners, drifting like ghosts in a tight school over a white sandy bottom, are nearly silent. The shallow water, with its reflective surface, gives them the opportunity to visually monitor the world above. The white bottom spotlights any creature coming from below. The dolphins rely on the parameters of the school — what scientists call the envelope — to scan to the left and to the right, to the front and to the back. When asleep, they can temporarily turn off all or most of their metabolically expensive hearing apparatus and rely on the collective eyes of the school. And so young and old, male and female, drift through sleep dependent upon their fellows, who in turn rely on them, to keep one and all safe.

Plutarch said that all men when awake are in one common world but that when asleep each is in a world of his own. This may not be true for dolphins, who, in their brightly lit beds, their eyes wide open and their ears closed, are almost certainly experiencing something different than we do in the five stages of human sleep. We don't know, for instance, if dolphins dream — since that window to the unconscious would surely prove lethal. If they don't dream, then what is happening in their minds as they float over the sandy bays of the tropical world?

Scientists ask, Why do we sleep?, seeking answers from biology, chemistry, psychiatry, psychology, and genetics. On a physiological level, we know that human sleep provides an anabolic environment wherein the body repairs many of its parts and stores energy. Wakefulness creates a catabolic environment that erodes and destroys many cellular processes while expending energy. Sleep is restorative. We can feel this, and apparently we need sleep more than we need wakefulness, since with the help of technology we can maintain sleep (coma) for years without awakening, while sleep deprivation quickly proves lethal.

Alternately, we might ask, Why do we awaken? If the state of wakefulness is not the true world but the one corrupted by our senses, our memories, and our expectations, perhaps the state of sleep is closer to the true world. The Tibetan Buddhist lama Tenzin Wangyal Rinpoche advises us to look to our experience of sleep to discover whether we are truly awake:

. . . every day ends the same. We shut our eyes and dissolve into darkness. We do so fearlessly, even as everything we know as "me" disappears. After a brief period, images arise and our sense of self arises with them. We exist again in the apparently limitless world of dream. Every night we participate in these most profound mysteries, moving from one dimension of experience to another, losing our sense of self and finding it again, and yet we take it all for granted. We wake in the morning and continue in "real" life, but in a sense we are still asleep and dreaming. The teachings tell us that we can continue in this deluded, dreamy state, day and night, or wake up to the truth.

The *Mandukya Upanishad,* one of the hymns of the Vedanta sages, describes the state of deep sleep as the third aspect of Self, where the mind rests with awareness suspended. "This state beyond duality — from which the waves of thinking emerge — is enjoyed by the enlightened as an ocean of silence and bliss."

Deep in Opunohu Bay, the spinners drifting over the sandy bottom inhabit their own silent ocean. Their bed is buoyant and weightless, blissfully free of bed springs, mattresses, pillows. Suspended in the ocean's womb, they enjoy a homeostatic world darkened to their ears and pleasantly transparent to their eyes. If there are predators nearby, the spinners, slumbering with their friends, will rally and drive them away. Sleep is the best meditation, says the Dalai Lama, and although the dolphins may not be dreaming, perhaps they are meditating, a state advanced human practioners substitute for sleep.

31

The Sleep of Plants

THE ʿATAE OUTSIDE my *fare* in Village Faimano, with its red flowers blooming on bare branches, and its noisy population of honeyeaters, undergoes a form of sleep, too. In the daytime, alert and awake, it displays what biologists call heliotropic (Greek *helios:* sun; *tropos:* turn) movement, as its green baby leaves, as soft as butterfly wings, demonstrate an intelligent orientation with regards to light. Each day the new leaves track the sun as it arcs overhead. In the cooler early morning and late afternoon hours they exhibit a diaheliotropic orientation, turning their flat surfaces perpendicular to the sun's rays for maximum exposure. In the hottest hours of midday, they shift to a paraheliotropic orientation, turning only their edges toward a sun so strong it would likely burn their photosynthesizing organelles.

Like the family of French tourists inhabiting the *fare* next door to me, the ʿatae leaves spread themselves out to soak up the kinder sun of morning and late afternoon, then hide as best they can at midday. At night, when the French lay their small children down on the cooler veranda to sleep, then sit in the dark drinking wine and smoking cigarettes and dreaming aloud for hours of their hopes to *parcourir les mers* (to sail the seas), the ʿatae sleeps, too. In the moonlight, the leaves relinquish their skyward stretch to hang upside-down, like a tribe of green bats in lacy caves of flowers.

This response by plants to darkness and/or temperature changes is known as nyctinasty (Greek *nyktos:* night; *nastos:* pressed closed). The mechanism by which the *'atae* changes its leaf orientation is its pulvinus, that padded joint between the stem and the leaf that acts as a motor organ. During sleep, the leaves change not only their position but also their function, beginning the "synthesis" stage of photosynthesis, through processes known as the dark reactions (or the light-independent reactions). Leaves drooping, the *'atae* diffuses carbon dioxide molecules through its skin, where they make contact with the plant's photosynthetic cells and begin a series of molecular reactions known as carbon fixation and the Calvin-Benson cycle. Together these reactions produce the glucose that feeds the plant.

Come morning, the leaves awaken, assume their diaheliotropic orientation, and begin to perform the light reactions (or light-dependent reactions). In this half of the photosynthetic cycle, the glucose made in the night is converted into energy through a process known as cellular respiration. Some of the waste products of this are the oxygen molecules stripped from water during photolysis (Greek *photos:* light; *lyo:* split up). These molecules are then diffused through the plant's skin and into the air.

As best we know, the dark and light reactions of photosynthesis arose about two billion years ago in photosynthesizing bacteria known as cyanophytes or blue-greens. This was when all life on earth was still aquatic, or so we currently think. And so for the next three hundred million to half a billion years, most or all of the newly released waste oxygen molecules dissolved into the seas, the lakes, and the muds of the primordial world. But by 1.5 billion years ago, some oxygen was diffusing into the atmosphere, where it radically transformed the recipe of the air and facilitated the emergence of life forms practicing aerobic respiration. In other words, the sleep of plants produced nothing less than the beginning of the age of animals.

Most nights at the Village Faimano I sleep on my veranda, willing the scant breezes on this leeward side of the island to come my way. A gibbous moon up these nights casts shadows through the 'atae separating my *fare* from the French family next door, providing an illusion of privacy while doing nothing to dampen sound. The French couple stays up late and chats to old friends on the island who stop by. By eavesdropping, willingly or unwillingly on these hot and humid nights when sleep is less a reality than a dream, I come to learn how they sailed here many years ago. Half-awake, half-translating, I can almost see their yacht, *Pétrel,* bobbing through Tareu Pass — the same pass the spinner dolphins exited only a few hours ago en route to adventures in the darkness of the abyss.

The spinners, like many if not most of the ocean's nocturnal feeders, are dependent on the mysterious movements of a vast community of the mesopelagic (Greek *mesos:* middle; *pelagos:* open sea), known as the deep scattering layer (DSL). This aggregation of life forms was unknown until the 1920s, when early hydrographers, mapping the ocean bottom with echosounders, encountered evidence of a sea floor that appeared to rise at night and fall by day. When the sun was up, the sea bottom seemed planted at around thirty-three hundred feet, then mysteriously migrated to less than six hundred feet in the darkness, sometimes to as little as thirty feet.

Although the DSL (named for its echosounding signature) was a source of confusion to hydrographers — who for a while wondered if all the world ocean was only three thousand feet deep — it fascinated marine biologists. Their investigations eventually revealed this mysterious layer to be composed of living creatures hiding inside the uppermost limits of permanent darkness during daytime and rising toward the surface at night. Continuing investigation revealed the DSL's species composition to be about 80 percent myctophids (lanternfishes) and euphausiids (krill).

The echosounders of old were actually reading the millions and millions of swim bladders of the myctophids — the two hundred forty-six known species of lanternfish who inhabit the dark world of the DSL. For the most part, these are ordinary-looking fish, between one and twelve inches long, endowed with the usual teleost hardware of fins, scales, lateral lines, and tails. But their nyctipelagic lifestyle — that is, their habit of hiding below the cusp of darkness by day and rising to the surface at night — led to the development of extraordinarily large eyes and a series of light-producing organs known as photophores.

Hiding within perpetual darkness, these lanternfish produce their own light, usually a weak blue, green, or yellowish, whose color and pattern impart information about the fish's species, its gender, its behavior, as well as information used in shoaling, and other communications we cannot decipher. Moreover, these photophores are generally arranged in rows along the fishes' flanks, creating a camouflage known as counterillumination. By adjusting their internal dimmer switch, the lanternfish use their photophores to match the slightest overhead ambient light levels — whether it be the faint light of the sun or the moon — making their silhouette less visible, maybe invisible, to predators below. A few lanternfish of the genus *Diaphus* possess photophore "head-lights" situated next to their eyes, which may be used to locate prey — though because these headlights are larger in males than in females, they likely play a role in social signaling.

The myctophids account for 65 percent of all mesopelagic fishes. All told, they comprise an estimated global biomass of as much as six hundred sixty million metric tons, giving them perhaps the greatest distribution, population, and species diversity of all vertebrate species on the planet. Yet lanternfish are not the only members of the DSL, which is also populated by many squids, jellyfishes, ctenophores, shrimp, copepods (small crustaceans), mysids (opossum shrimp), arrowworms (planktonic worms), and pteropods (sea butterflies). These smaller species make up a significant portion of the mesopelagic zooplankton, upon which the lanternfish feed heavily or even exclusively.

It's no wonder then that many if not most of the predators of the epipelagic food web — the surface-dwelling salmons, tunas, billfishes, manta rays, pelagic sharks, whale sharks, basking sharks, dolphins, whales, seals, and seabirds such as penguins — make their living diving to meet the rising DSL. Each dusk, it's as if a great dumbwaiter rises from the deep bearing every manner of seafood delicacy on a platter of darkness. The challenge for the surface hunters, blindfolded by night, is to locate the components of this seafood platter, to catch them and eat them without befalling the same fate themselves.

32

Living Lanterns

SOME SPECIES OF THE deep scattering layer rise at night to avoid the daytime hunters such as tuna. Others rise at night to avoid the nocturnal hunters of the mesopelagic — the lancetfish, for instance, who do not perform a vertical migration. Others hunker down in the mesopelagic during the day as an energy-saving measure, allowing their bodies to enter a suspended animation, or sleep, in the cold water below the thermocline.

Ecologists have proven that it's more efficient for an organism to sleep at three thousand feet during the day, swim the human equivalent of twenty-five miles upward, then work all night and swim twenty-five miles back to bed. The alternative, living within warmer surface waters around the clock, is less efficient, producing a faster metabolism requiring more food to fuel it. Backing up this supposition is the fact that waters lacking a temperature differential between the surface and the deep — for example, the extreme Arctic and Antarctic — do not contain a deep scattering layer.

Other denizens of the DSL rise at night and sink during the day to avoid being swept off to distant latitudes or climate zones. Copepods in the Southern Ocean ride surface currents in one direction all night, then sink into a mesopelagic countercurrent that returns them to near their starting point during the day. Others inhabiting coastal areas with strong upwellings follow surface drifts offshore during the night, then ride deep inshore flows back during the day.

The most abundant invertebrates of the DSL, the euphausiids, or krill, are probably conserving energy down deep, as well as riding the currents, to stay in place. But they are also certainly rising at night to graze on the pastures of the sea. Most krill are omnivores and filter feeders that use their comblike "legs," known as thoracopods, to sieve plankton from the water. Among their most abundant prey are the phytoplankton — the single-celled plants that can survive only in the sunlight zone. To reach them, the krill must migrate upward.

The fantastically tiny unicellular realm of the phytoplankton contains some of the most beautiful life forms on our planet, including the ethereally geometric diatoms, the whip-bearing dinoflagellates, and their even tinier counterparts, the nanoplankton and the picoplankton (which are sometimes collectively called the nanoplankton, from Greek *nanos*, dwarf). This microcosmos burgeoned enormously recently with the discovery of the prochlorophytes: phytoplankters measuring three hundred-thousandths of an inch, which reach population densities of thirty million cells per ounce of seawater, and which may account for as much as four-fifths of all the photosynthesis in the world ocean.

Dependent on the sun, none of these phytoplankton, whatever their size, survive below the epipelagic zone. Yet many perform their own shallow version of a vertical migration, sinking to sixty feet deep at noon, then rising to within ten feet of the surface at night. In the darkness, performing the synthesis stage of photosynthesis, they hide close to the clutter of waves, avoiding the rising predators of the DSL.

Because each evening the krill are likewise rising — in their case from as deep as three thousand feet, scrambling upward on their swimmerets, or swimming legs. They are glowing softly, producing bioluminescence from their own onboard photophores.

The extent of krill predation on dinoflagellates can be construed from the fact that the krill's light-producing pigment, known as luciferin (Latin *lucis:* light), is nearly chemically identical to the bioluminescence produced by dinoflagellates. Apparently krill do not produce their own glow but ingest it from their meals.

Accompanying the krill upward are the lanternfish. Hiding within their own glow, they feed on the krill, which are feeding on the bioluminescent plants of the sea. Diving to meet them are the surface predators — including the sharks, billfish, and dolphins — most of whom, although they do not glow of their own accord, trigger a fireworks show as they sprint and lunge and mechanically activate the bioluminescence of thirty billion single-celled plants inhabiting every cubic foot of surface seawater.

Years ago, during a winter of intense El Niño storms, I was part of a film crew sailing from San Francisco to Mexico's Sea of Cortez. Riding up and down the uncharacteristically large seas rolling out of the southwest, our ketch climbed to the summits of the waves, affording us a lookout in all directions, then skewed down the backsides into the troughs, where the only view was of the walls of water ahead and behind.

Even in the night, the broad, rounded crowns of the swells afforded a vantage for miles across what looked like marching armies of waves, many of them breaking in a tumble of whitewater that ignited all the bioluminescent organisms within. Each breaking wave flared intensely blue-green and then went dark again, as if an underwater explosion had taken place. This panorama was akin to a mysterious messaging system, as windows of waters flashed luminous colors, then blinked shut, the whole spectacle animating countless miles of sea like a marine semaphore on a monumental scale.

Adding to these distant vistas was the vision in the trough between the swells, where an incandescent waterfall of radiance tumbled from the wave breaking directly ahead. Time and again, we sailed into this cascade, which outlined the bow of the sailboat in a vibrating aura of electric blue and electric green. Behind us, our wake writhed like an aquamarine serpent before fading from view over the precipice of the receding wave.

Most nights schools of dolphins, generally common dolphins (*Delphinus delphis*), arrived seemingly out of nowhere to ride our bow wave. Alone on watch in the open-air cockpit, I would glimpse a blueish-green streak heading toward the bow, looking for all the world like a torpedo, until it porpoised through the surface to breathe. The first streak was invariably followed by others, sometimes dozens, all racing across the beam, carving glowing turns before settling into bow-riding position: on their sides, tail flukes nearly touching the bow as it hobbyhorsed through the swells. No matter how violent the boat's action, the dolphins held fast in their position, their bodies outlined in the pulsing blue light of an underwater Saint Elmo's fire, their whistles audible through the surface, even above the hissing of the breaking waves, the cleaving of the bow, the clanking of piston hanks, and the creaking of a boat head-on to the seas.

Common dolphins were not the only cetaceans we encountered in those bioluminescent nights. We also came upon migrating gray whales headed for their breeding lagoons on the western coast of the Baja Peninsula, whose bioluminescent wakes mimicked the wakes of boats. If conditions were right, when the whales blew, the plugs of water covering their blowholes flared into faint, blue-green mists of luminous organisms, as if the behemoths were exhaling pixie dust.

Although 90 percent of the creatures of the mesopelagic deep scattering layer are capable of bioluminescence, the light show we observe near the surface, at least as we understand it, is produced largely by the dinoflagellates — those unicellular plants possessing microscopic whiplike tails that enable them to move, however slightly, this way and that. Large creatures such as you or me or a spinner dolphin will activate billions of these bioluminescent plants as we move through the surface layer of the nighttime waters. Even small zooplankton, for instance, the predatory krill traveling on their swimmerets, will find themselves spotlighted.

As to why plants in the sea produce bioluminescence, a hypothesis known as the burglar alarm theory postulates that the chemical production of light acts as a visual siren: the plants turn on their lights when their predators (for example, krill) are in motion, illuminating them so that *their* predators (for example, lanternfish) can catch the krill and eat them first. Although useful as such, in all likelihood the burglar alarm is a serendipitous side effect of the original purpose of bioluminescence, which most likely evolved as a chemical defense against oxygen.

This original purpose dates back to the emergence of the first oxygen produced by early photosynthesizers. Since none of the unicellular life on earth at that time possessed any protection from oxygen's corrosive properties, many perished. But a few enterprising species concocted a recipe still followed by today's bioluminescent organisms: an enzyme of luciferase, combined with a molecule of luceferin, stirred with a dash of oxygen and salt to produce a flash of light that consumes the oxygen molecule. Although modern plants no longer require bioluminescence to survive, since they combat oxygen through respiration, their ancestors were likely saved by this effervescent cocktail of the sea.

33

The Clamor of
True Democracy

THIS AFTERNOON, in Opunohu Bay, the spinners
are beginning to awaken. From my vantage in the
va'a, the first clue is auditory rather than visual, since
I am looking the wrong way. Hearing the telltale slaps at the surface, I turn my canoe and see two spinners rising beak first,
smacking their snouts sideways on the surface. They are exhibiting the behavior scientists call the nose-out: lunging upward,
slender beaks breaking the surface before rolling over and diving
again. One spinner's nose-out is accompanied by a slap, as it
snaps its rostrum (Latin *rostrum:* snout) against the surface. The
whole school is beginning to move more quickly now, as the dolphins shake off their somnolence.

On the next surfacing, a half-dozen nose-outs are quickly followed by the more dramatic aerial displays for which spinners are
named. Many begin to perform what researchers call salmon
leaps: launching headfirst out of the water, arching the back and
flaring the pectoral fins before falling stiffly to the surface on the
side or on the back in loud, splashy flops. Other dolphins begin to
spin: launching headfirst out of the water while pirouetting rapidly on the axis and twisting through the air, before landing on
the surface in a belly flop or a back flop.

Most spinners perform a series of spins, one after another, up
to fourteen of them at a time, with each spin growing smaller
and less intense as the dolphin wears itself out. By my count,

from editing slow-motion footage, the most accomplished dolphins can execute 7.5 revolutions on their axes during the 1.25 seconds they are in the air. The purpose of these displays is communicatory, and the emphasis is not on the topside spins and twirls that so delight us, but on the underwater bubble plumes formed from the splash. These are far more "visible" to the dolphins' hearing than anything their eyes perceive, and enable the school to maintain contact even when widely dispersed.

Spinning is the defining art of this species, and even infant dolphins practice it nearly from the moment of birth. It's not uncommon to see tiny newborns hugging their mother's flanks, then flipping themselves out of the water and falling onto their sides. Spins are the sole domain of *Stenella longirostris* and its sister species, the short-snouted spinner dolphin, *Stenella clymene*. No other dolphin species, no matter how aerially accomplished, performs spins or can be trained to perform spins.

In the late afternoon, as the sun descends behind Mount Tautuapae, driving a long shadow across the inner reach of Opunohu Bay, the airborne dolphins appear as tiny black figures against a monolith of towering peaks, the splashes of their reentries opening and collapsing around them like lace petals. They are moving fast enough now, with members of the school popping out of the water in all manner of leaps and spins, that I no longer try to keep up, and watch from afar as they race for Tareu Pass.

Then, as quickly as they depart, they return — not quite to the white bed of the sandy bottom, but to a shallow area nearby, where they appear to fall asleep again. This vacillation between sleep and wakefulness marks the stage of their day that scientists call zigzag swimming. As the subgroups draw closer and resume their cryptic breathing, the school once again rises and falls in ghostly unison, although a few individuals appear restless and continue to head-slap, tail-slap, and leap. Their splashy entreaties persist for the next thirty minutes or so, until they succeed in convincing a few of their fellows to awaken, then a few more, until once again the whole school is leaping and slapping as they pick up speed and head toward the pass.

On this second trip across the bay, the numbers of leapers increases as the school spreads out, enlarging the size of the envelope. The school's speed also increases, along with the number of aerial behaviors, until the dolphins cover an area of the bay three times larger than their sleeping school. The envelope of the school now bursts with spins and salmon leaps, along with dramatic tail-over-head leaps, as a dolphin exits the water headfirst, flips its tail over its head in a backward somersault, and reenters tailfirst, usually on its back with a sizeable splash. Some animals combine the tail-over-head with a spin to create a manic corkscrew that carries them in unpredictable directions, sometimes in reverse of the school's motion of travel.

Punctuated with spins and leaps, the envelope becomes wildly three-dimensional, even from a topside perspective, and just when it seems the spinners will actually spin themselves out Tareu Pass and into the open sea, they turn around, close ranks, and drift back toward the inner bay. Once again their aerial behaviors subside as the dolphins conceal themselves within the clutter of waves and against a sandy bottom. For the next forty-five minutes they resume sleep.

The nondreaming parts of human sleep are spent consolidating and optimizing memories, including a critical process known as unlearning, in which we erase unwanted thoughts and erroneous information. Unlearning culls the amount of data the brain must maintain and manage, freeing us to create new memories and to assimilate new knowledge. Sleep therefore is an opportunity to dismantle the cognitive scaffolding that props up the world we know as wakefulness. And just as every day we build new scaffolding from the supple bamboo of memory and association, every night we tear some of it down, lest we build ourselves into a cage.

The Bön tradition of Tibetan Buddhism teaches a yoga of dreams and sleep wherein, among other things, practitioners learn to consciously dismantle the scaffolding. This work is done in the course of lucid dreaming, which students learn to cultivate at will. Since dreams are free of the rational mind, they reveal our true level of awareness and can provide an alternate and possibly speedier pathway to a clearer life. Although Western psychology believes that dreams should not be tampered with, since they carry messages from the subconscious, Buddhists think differently. According to Tenzin Wangyal Rinpoche, "it is better for the aware dreamer to control the dream than for the dreamer to be dreamed."

Late this afternoon in Opunohu Bay, dreaming or not, the dolphins are caught in the pendulum of zigzag swimming — swinging between a state of sleep and a state of alertness, between the comforts of bed and the thrills and rewards of the waking world. There is no equivalent in human life, and the dolphins' adeptness at managing these changes suggests a mastery not only of a lucid sleep but, perhaps, of a lucid wakefulness, too.

Individual spinners do not awaken alone at the behest of an individual clock, internal or external. The process of awakening, as with everything in their lives, is arrived at by group consensus. As the pendulum swings between wakefulness and sleep, some dolphins are eager to awaken and others are desirous of rest. The eager ones try to energize the group with aerial displays, perhaps because they are hungry and anxious to get out to sea where all the food awaits, or perhaps because they fed well the night before and feel livelier now. Whatever their motivations, the energies of the restless individuals are countered by the energies of the sleepy individuals, who prefer to lounge in bed a while longer.

The dynamics of synchronization, which are obvious from topside, are even more so from below, although this is nearly impossible to observe among Tahitian spinners who are unhabituated to human swimmers and will vacate the area when a human snorkels nearby. And, so, dropping a hydrophone, I eavesdrop on the debate. It's obvious when the sleepers win the vote, since the school falls silent. It's also obvious when the wakers begin to canvass, as the headset crackles with whistles, clicks, and the characteristic quacks, moos, baahs, barks, and squawks of the burst-pulsed signals. In short order, these sounds are accompanied by a series — sometimes an artillery barrage — of dull explosions and hissing bubble-trains: the percussions of belly flops and back flops at the surface.

Few encounters are more sublime than floating on a *va'a*, rocked by the ripples on the lagoon, eyes closed, ears open to the world below. The air is thick with the moisture of the sea, with the sharp scent of the afternoon thunderstorms on the summit of Mount Tohiea, and with the fishy smell of the spinners' breath, like a spritz of mackerel. Cradled in your canoe, rocking in place with your eyes closed, experience collapses into a bubble, transparent to a larger world yet self-contained. When the dolphins awaken, the bubble bursts, exploding your senses into the growing dimensions of the school, stretching the envelope of your awareness, so that even without looking you can see through your ears the growing consensus and excitement of the spinners' dash to the sea.

If there is another sound underwater at this moment you are unlikely to hear it behind the cacophony of dozens of spinners psyching themselves for the race offshore. At this moment, the whole world is their barnyard orchestra, their penny whistles, the electric buzz of their echolocation, the timpani of their leaps. Theirs are not the archetypal voices of humpback whales, but a stream of SETI data: the sound of extraterrestrial life, full of meaning unintelligible to us.

Consolidating their intentions, the spinners' voices grow louder and merge into the congested crosstalk that Norris and his colleagues jokingly coined the Yugoslavian News Broadcast. This is the buoyant clamor of true democracy. Since there is no leader or hierarchy in this, or any other, aspect of spinner life, every dolphin sleeping in this bay is awarded the same voting power. However many spinners reside here today is the same number that must now reach consensus on when to leave and where to go.

The underwater sight of spinner dolphins dashing to the open sea is a marvel of speed and grace. Dozens of slender bodies stream below the surface at velocities that transform the school into a waving contrail of gray and white and black. The individuals space themselves with the jostling precision of fighter jets, sliding in and out of position with controlled yet efficient effort. Most animals array themselves at the level of their neighbors' pectoral fins, creating an echelon that maximizes the flow field between individuals and minimizes drag in the water. Young animals draft directly alongside adults, riding just to the side, and just below an elder's pectoral fin, where they capture a discounted or even a free ride.

Moving thus at speed, the dolphins' bodies ripple as their flukes oscillate faster and at decreased amplitude so that the motion of the tails appears to blur. In fact, the school's movement as a whole has an out-of-focus quality, as if the cameras of our eyes are too slow to capture their definitions. Moving fast — though not yet at top speed — the school travels twenty feet below the surface, at a depth where the propulsive upthrust of the tail flukes avoids the energy-sapping interference of the surface.

Seen from above, this interference appears as the characteristic whale or dolphin "footprint," the circular pool of water surrounded by raised ridges that marks each upthrust of the flukes. Seen from below, these footprints appear as tiny whirlpools of swirling bubbles that boil to the surface, creating flat, glassy windows to the topside world. Whale footprints, often ten feet wide and spaced fifty or more feet apart, enable the human observer to track an individual's travel even when it's swimming deep underwater. The footprints of a fast-moving dolphin school tend to merge into series of connected footprints something like a wake.

Years ago, while making a film about spinner dolphins, I was part of a crew that shot most of the underwater footage twenty-eight hundred miles north of Mo'orea in the Hawaiian Islands. Thanks to the greater overall number of boaters, swimmers, snorkelers, and divers in Hawaii, many spinners there are at least somewhat habituated to humans.

In the late afternoon, the Hawaiian dolphins awakened and began the series of votes comprising zigzag swimming — dashing away from the protection of shallow water toward the Kealaikahiki (Hawaiian for "the way to Tahiti") Channel. But they did not get far before they were persuaded by those still sleepy, or by those who believed it was too early in the day to set to sea, and the school drifted inshore and milled itself back to sleep.

When the dolphins awakened again and initiated another dash, we in the film crew raced ahead, gunning the Boston whaler to what we imagined was an intercept point on the spinners' course. Cutting the motor, we slipped overboard, cameras in hand, and waited. Since we could not hope to intercept a fast-moving school except by direct hit, there was no point in swimming to the left, if that's where the school turned, or swimming after them, if they reversed course. Time and again, by trial and error, we sped off, jumped overboard, waited, and usually missed. It was exhausting work, hauling bodies and cameras in bulky underwater housings in and out of the boat, often in heavy seas. But when it worked, it awarded us that rare thirty- or sixty-second run of footage of a contrail of streaming dolphins.

For a moment — stationary in the water as the echelons of spinners streamed past, spaced in their graceful geometry — we had the feeling of being inside the eye of a hurricane. The blurry contrail was embroidered with the underwater bubble plumes of dolphins leaping and spinning and landing on the surface. The sight alone was spellbinding. Combined with the auditory explosion of the Yugoslavian News Broadcast, the experience was so contagiously exciting that I doubt there was one among us who would not have swum off with the spinners into the sunset if we could have physically managed it.

34

The Spirit of Godly Gamesomeness

RESEARCHER KEN NORRIS and colleagues have noted the similarity of spinner dolphins to other social species who perform complex social facilitation rituals. These are group activities, something like pep rallies, designed to synchronize the group's intentions. African wild dogs sing to unite the group before the hunt. European jackdaws (raven relatives), use croaking choruses to debate on where the group should roost for the night.

We are also members of a social species whose decisions are likewise influenced by the group, subtly or grossly, but probably to a greater extent than we are aware of or admit. Because of this, the draw of social-facilitation rituals is palpable to us, even across species boundaries. Biologist Bernd Heinrich speaks of this when he writes:

> There is something in the ceaseless chatter of migrating geese that stirs me. Perhaps it touches something wild, remote, and mysterious that I share with them, for it is almost with longing that I look up each fall and spring when the scraggly V formations wing their way overhead high in the sky.

Even with spinner dolphins, in a world as alien to us as another planet, we are influenced and persuaded by the call to mobilize — though we cannot hope to emulate those whose talents, skills, sensory abilities, and endurance exist on a scale far beyond our capabilities, even our technological capabilities. So we follow the spinners by boat as best we can, while the school commits to its outward course. Once outside Mo'orea's barrier reef, the spinners jointly gun their accelerators until they pass the threshold known as the crossover speed. This is the point at which it becomes more efficient for the dolphins to travel in the air than in the water.

The school is suddenly bounding out of the oceanic swells in arcuate leaps. These are the typical porpoising jumps that all dolphins and most, if not all, whales perform: headfirst out of the water, the body arcing two or more body-lengths in the dry world before entering again headfirst. Unlike the percussive salmon leaps and spins, arcuate leaps are designed for speed and ease of travel and hence are nearly splashless.

From underwater, the mechanics are clear. The dolphins swim below the cascade of bubbles formed by the breaking tops of the oceanic swells then angle upward, beating furiously with their flukes to reach top speed before entering the dry world. For a moment, their bodies enter relatively frictionless air, while their tails thrust once more underwater. When even the tail is airborne, the dolphins begin flying. They soar forward, racking up far more distance and speed than they ever could swimming solely underwater.

Yet even as the school is traveling fast and hard in graceful arcuate leaps, members are still spinning and corkscrewing across the envelope. There is hardly another sight on our ocean planet like that of the tiny spinners suddenly bursting out of the backside of a swell so big it dwarfs them, the school swinging in unison on a trapeze of air, while some members explode into exuberant spins that carry them backward or sideways, like drunken acrobats. Herman Melville, in *Moby-Dick*, dubbed them the huzza porpoises because

[they swim] in hilarious shoals, which upon the broad sea keep tossing themselves to heaven like caps in a Fourth-of-July crowd.

Their appearance is generally hailed with delight by the mariner. Full of fine spirits, they invariably come from the breezy billows to windward. They are the lads that always live before the wind. They are accounted a lucky omen. If you yourself can withstand three cheers at beholding these vivacious fish, then heaven help ye; the spirit of godly gamesomeness is not in ye.

And yet, as hilarious as these shoals are, as infectious of spirit, the purpose behind such behaviors is deadly serious. In setting out to sea, these diminutive dolphins face the most dangerous and demanding part of their lives. It's likely that their pep rally instills courage as well as group cohesion, since the spinners are headed into the darkness, where they will be lit with bioluminescence, and where their chemical, flow-field, and electrical signatures will be available for every predator, large and small, to see.

No human has ever seen spinner dolphins working the depths of the deep scattering layer at night — although researchers have followed a few individuals via radio tracking equipment to dives of one thousand feet or more. The last we see of them, tracking by boat in the twilight offshore, is a glimpse of the enigmatic behavior that Norris and colleagues described as the spread formation. The racing dolphins suddenly stop. The school breaks into subgroups and spreads across miles of ocean. "The observer," writes Norris and others, "who had moments before been traveling with a coherent school may look up in bewilderment, wondering where the school went."

This is the moment where travel transitions to hunting, as the members disperse to look for food. Diving alone or in small subgroups, perhaps no bigger than mother-calf pairs, the spinners make deep penetrations from the surface into the abyss. They are diving hard, descending past the bioluminescent surface layer into the darkness in search of the mesopelagic layers where lanternfish, barracudina, blacksmelt, squid, and other prey congregate by species in discreet strata. Despite being physically separated, the dolphins are nevertheless still hunting cooperatively, tethered to one another by the sounds they make, and by the sounds they hear. As soon as a viable prey stratum — squid or lanternfish — is located, the school cinches the envelope closed, as all the members coalesce to feed inside the safety net of their joint defenses.

But reconvening in the dark is no mean feat, as evinced by the spinners' soundtrack, which crescendos into an overload of burst-pulsed signals, whistles, and echolocational clicks. The whistles amplify in the dark, building themselves into what Norris and Shannon Brownlee describe as a "rhythmic series producing interweaving tangles of whistles that create an ululating effect." Heard from topside, it's a chilling soundtrack. Underwater, the schools of lanternfish and squid, rising in pursuit of krill, are bathed in its bansheelike wailing.

The one hundred twenty-three books of the *Upanishads,* the classical Hindu texts on meditation and philosophy, have come down to us in Sanskrit — the oldest of the Indo-European languages, and the mother tongue for virtually all European languages. Yet the *Upanishads* date much earlier, to an indeterminate span of time when they were transmitted orally in what translator Alistair Shearer calls the "sacred language par excellence, the Vedic." According to Shearer, this ancient language was believed to be

> not merely a conventional system of representation based on linear logic but the language of nature herself, composed of the primordial sounds that promote order in the evolving universe. These sounds, like music, communicate preverbally and have a universal meaning that transcends all cultural boundaries.

Among these primordial sounds is that most ancient of mantras, Om, whose vibration is said to continuously generate the universe. According to the *Mandukya Upanishad,* Om consists of three constituent parts, A-U-M. The *A* corresponds to the outer world and our waking state; the *U* to the inner world and our state of dreaming; and the M to the state of dreamless sleep — what the *Mandukya Upanishad* calls the ocean of silence and bliss.

The notion of sound as a generator of the universe seems easier to comprehend in the underwater world than in the dry one. The echolocational click-trains of dolphins, for example, generate sounds that draw pictures of the world around them. These sound-pictures are available to dolphins both in darkness and in light and can penetrate substances to provide an x-ray–type image of the insides of the world. Furthermore, dolphin sound-pictures are transmissible to others in the school — as if we could instantaneously print a photograph of what we see with our eyes and send it to any person within hearing range, which is to say a long way away underwater.

The extent to which sound binds dolphin schools together was first put forth in 1980, when Norris and Tom Dohl proposed that all members of a spinner school share information through what they called a sensory integration system, or SIS. Each dolphin acts as one component of a supraindividual signalling system, akin to one pixel on a computer screen, and each member can tap into anything and everything that every other member of the school is learning, seeing, hearing, or feeling to create a much larger and fuller picture of the world around them.

Norris and Dohl called these combined informational pixels the collective sensory windows of the school, and theorized that they allow dolphins to see not only through their own sensory window but through everyone else's as well. Thanks to the properties of sound propagation underwater, information within the SIS travels fast, far, and accurately, enabling dispersed members to coordinate their operations within the school as a whole.

Thus the SIS generates the dolphins' universe, enveloping them in a world of meaning that affects not only their response to their environment but the parameters of the environment itself. For instance, using an echolocational click-train, a dolphin diving into the DSL might discover a school of lanternfish and transmit information about it to the other spinners via sound-pictures. But then she might also direct an intense burst of sound to disable the fish, causing one or more lanternfish to wobble and disrupt the schooling structure. Other spinners arrive to partake of the debilitated prey. In this way, the signaling dolphin uses sound to observe her world, to disseminate information about her world, which then changes the behavior of her school members, as well as the behavior of the fish school. As the dolphins feed, the fish die, and the universe is made.

Overall, the phonation abilities of odontocetes (Greek *odontos*: tooth; Latin *cetus*: whale) are among the most remarkable known to us. Researchers recorded bottlenose dolphins (*Tursiops truncatus*) emitting echolocational clicks of two hundred twenty-nine decibels intensity, the outermost limit of sound, beyond which additional energy will result in the production of heat, not more sound. In 1983 Norris and Bertel Møhl proposed that these ultra-intense sounds were used for prey stunning — a behavior since observed in bottlenose dolphins and in Atlantic spotted dolphins (*Stenella frontalis*), among other cetaceans.

Along with dolphins, sperm whales (*Physeter macrocephalus*) also produce sounds loud enough to stun fish and squid, and this is likely their primary food-capture method, since whales with broken and mangled jaws manage to survive and are otherwise healthy. The most remarkable adaptation of all is seen in males of an elusive species of beaked whale known as the strap-toothed whale (*Mesoplodon layardii*), who grow a single tusklike tooth from each side of their lower jaw that arches and overlaps above the upper jaw. Thus completely muzzled, they cannot open their mouths more than an inch or two, and presumably rely on prey-stunning sounds to immobilize their food, which they vacuum into their mouths with specialized tongues.

As to how ensonification immobilizes fish prey, Norris and Møhl theorized that the intense bursts of sound produced by toothed whales and dolphins disables the fluid-filled inner ear, or labyrinth, of teleost fishes. Since the labyrinth provides gravitational orientation (pitch, roll, and yaw), fish at the receiving end of an ensonification blast suffer the underwater equivalent of vertigo.

Dizziness in the sea is effectively a death knell. Most fish scales are embedded with reflective guanine molecules designed to camouflage by reflection. When oriented, a fish's scales mirror the world, making the fish less visible when seen from above, from the sides, or from below. But if a fish loses its vertical orientation and wobbles, then the guanine mirrors begin to flash random reflections — the underwater equivalent of an SOS. Even if ensonification occurs at night, dizziness betrays the fish through disruptions in the flow field. As a result, maimed fish struggle to maintain their vertical orientation until nearly the moment of death, because the loss of orientation means certain death in the jaws of one or another of the sea's ever-ready predators.

35
Coral Noose

FROM THE COBALT waters beyond the barrier reef, the island of Mo'orea is breathtaking: the sharktooth ridges, the spires, the verdant valleys bathed in the turquoise moat of the lagoon. The ocean's big waves rise onto the barrier reef and break in white combers that toss their spray back at us, scenting the air with the atomized fragrance of the deep. We are observing all this from the bow of the *Mareva P* — a twenty-foot, twin-hulled vessel designed to carry dolphin-watching tourists on half-day and full-day tours of Mo'orea — while her owner, biologist Michael Poole, pursues his research on spinner dolphins.

On a day like this, with its complex of clouds and light and angles, it's easy to forget the mission that has brought us here. Through one pass after another we enter Mo'orea's bays and idle for ten to thirty minutes — whatever is required to scan for schools of sleeping spinners. Michael, an American with dual French citizenship and a resident of Mo'orea since 1987, knows every inch of every bay of the island, including which areas the dolphins frequent and which they avoid. He also recognizes many individual spinners by sight, based on identifying scars or nicks on dorsal fins, pectoral fins, and flukes.

We are looking for an old friend named No Tip, the twelfth spinner Michael named when he began his field study on the life of Tahitian spinner dolphins. She is one of his all-time favorite dolphins, he says, whom he has known longer than his own wife. Every three years, without fail, she brings her newborn infant close to Michael's boat in way of introduction.

If we can find No Tip, then we will muster some hope for the tragedy shaping up behind nearby Avamotu Pass. But so far there has been no sight of her, even though we have found sleeping spinners behind three passes, and even though No Tip's truncated dorsal fin — the result of a shark bite long ago — is easily identifiable in any group. Michael, normally energetic and cheerful, is subdued, and his thin face, creased from years in the sun on the tropical sea, wears an uncharacteristic weariness. He continues to rattle off facts about everything in the nearby universe — the height of the mountains, the names of their demigods, the fate of the pizzeria in Ha'apiti, the expensive new subdivision by the Beachcomber Hotel — but we are not fooled. The two of us on this miniature film crew have known Michael and his dolphins for years, and neither he nor the spinners are normal this year.

Four hours after launching our circle-island reconnaissance, we return to Avamotu Bay, having located Mo'orea's entire resident population of spinners without finding No Tip. Three weeks ago, she and her calf, along with twenty-eight other dolphins, slipped through the tiny passage of Avamotu Pass, a shallow doorway barely fifteen feet wide and framed by a brutal conglomerate of broken coral and algal ridge. Ever since that fateful morning, the pass has been closed by heavy surf breaking along its entire width, effectively trapping the dolphins inside. One week ago, No Tip and her calf disappeared from sight. Michael is hoping they have rescued themselves by slipping out the pass during the slight lull between the big sets of swells. But he has yet to see her on the outside.

The *Mareva P* idles inside Avamotu Bay, and we count how many of the original thirty dolphins remain, stopping at twenty or twenty-two — the same as yesterday. All told, eight of the original thirty ensnared dolphins have disappeared, and the dwindling few remaining in the bay are behaving unlike normal spinners, moving lethargically, holding motionless at the surface for many breaths, rarely diving to the bottom. Their typical sleep behavior has been disrupted, presumably by stress, and they are not moving in unison but in mother-calf pairs, or simply solo. The school they had depended on for survival has failed them, and they appear to have abandoned it. One youngster lingers only ten feet from our boat, a thin stream of bubbles leaking from his blowhole as he repeatedly emits his lonely signature whistle — a sound unique to each dolphin and one used to initiate conversation.

Avamotu Bay is an area of sandy bottom about the size of a football field, enclosed on the ocean side by the barrier reef, and enclosed inside the lagoon on all sides by a fringing reef rising to within three feet of the surface. Spinner dolphins, by their nature, will not traverse such shallow water — even though salvation lies a mere twenty yards away, on the other side of a narrow neck of this fringing reef. If these trapped dolphins could muster the courage or the audacity to cross it, they would be delivered into the far more spacious surrounds of Matauvau Bay, whose pass remains open despite the big surf. But day after day has passed without the spinners managing to surmount a fear as real to them as our fear of fire or heights.

The trapped dolphins appear thinner and more exhausted than yesterday, and their ribs are showing, something rarely seen in wild dolphins. Although reef fish abound in the lagoon, these are not part of the spinners' menu. Deprived of food for three weeks now, these dolphins are suffering less from starvation than from dehydration, since all their liquid intake comes from what they eat. They are also likely hypothermic. As we watch, one spinner repeatedly rolls upside-down, her beak breaking the surface at a forty-five-degree angle. Sadly, this is not part of the incredibly varied and exuberant social repertoire of her species, but the terminal distress of a water-dwelling, air-breathing mammal.

A few days earlier a spinner uncharacteristically spyhopped, head vertically out of the water, for five unbroken minutes as the *Mareva P* approached to within fifteen feet. When he finally slipped below, everyone aboard realized it was probably for the last time.

Three weeks into their entrapment in Avamotu Bay, the spinners are losing their vertical orientation, and although they do not possess reflective guanine molecules in their skin to signal their death throes, they are nevertheless flooding the near-field and the symmetrical flow field with the telltale reverberations of struggle. As if to confirm this, a very large lemon shark — an eight- or nine-foot monster, probably a female — prowls the edges of Avamotu Bay. It's rare to find a lemon shark inside Mo'orea's lagoon, particularly during the daylight hours — although they are certainly resident here — and neither Michael nor his boat driver, Heimata, has seen one inside before.

Heimata plays the outboards like a whisper, trying to guard the dolphins, while occasionally nudging the *Mareva P* back into the bay when the outflowing current tows us toward the white water of the pass. The spinners have split into two loose congregations today, one inhabiting the turbulent water near the pass, the other lingering far back in the bay, close to shore. Perhaps this division is indicative of a split in their hopes, with some imagining escape through the barrier reef and others imagining release across the fringing reef into Matauvau Bay. Or perhaps it is a response to the shark.

Because we feel useless and helpless we count the dolphins over and over again, and each time we come up with eight animals in the group at the back of the bay and twelve animals in the group near the pass. Michael talks quietly into his voice recorder, describing by rote the actions of the spinners as they endure this strange fate. His words are familiar, but his tone discloses the sadness of a man who must juggle the conflict between a professional ethic that discourages emotionalism of any kind, and a personal emotionalism that discourages indifference of any kind.

Avamotu is such a small pass that it doesn't even appear on most maps of Mo'orea, only on the boating charts, where it is punctuated with the nautical equivalent of exclamation points: a peppering of little black crosses marking dangerously shallow, rocky shoals. In all the years Michael has been circling this island and observing these resident spinners, he has found the dolphins inside this particular pass only once before — although he has heard of another entrapment twenty-five or thirty years ago.

He last found spinners in Avamotu exactly a year ago. Then, as now, Michael theorized they were chased through this bottleneck by a tiger shark, and that when faced with a passage between the shark's teeth or the coral shoals, they chose the latter. During that entrapment, the weather was more cooperative. Eventually the south-southwest swell closing out the pass moderated enough for Michael to mobilize seventeen boats to pull a *hukilau*, a Hawaiian-style purse seine net whose lines dangle vertically every six feet.

Since the *hukilau* contained no webbing, it could snare the dolphins only psychologically, and through these gentle means Michael hoped to force the spinners to commit to that which terrified them: crossing through. Michael's hope was that if forced to choose between the *hukilau* lines (six feet wide) and the pass (fifteen feet wide), the spinners would choose the pass. Some did and survived.

But the weather is not cooperating this year, and the pass is too closed-out with waves for the spinners even to consider it an option, no matter how much we might drive them toward it. Michael tosses out other ideas. Maybe he could use a sling to capture the dolphins, one by one, and carry them into Matauvau Bay. But spinners are notoriously fragile beings and have died from lesser interventions. Plus, even if he could manage to relocate them safely, what are their chances of surviving the energetic and tremendously dangerous flip-side of their life — that long trip out to the deep and the one-thousand-foot dives into the abyss? It's hard for us to imagine that the twenty wraiths in this bay are substantial enough anymore.

For air-breathing animals, dolphin or human, death in the water is caused by suffocation when the victim can no longer, for whatever reason, make it to the surface to breathe. Sinking and helpless, he or she will typically hold the breath in a voluntary apnea known as the mammalian diving reflex (MDR), named after the whales, dolphins, and seals for whom breath-holding is a fact of life.

The MDR provides a tiny window of salvation as the drowning body slams itself into preservation mode. The heart slows by 50 percent or more, a condition known as bradycardia, and is accompanied by peripheral vasoconstriction, as blood and oxygen are rerouted from the extremities to the vital organs, particularly to the brain. A blood shift also inflates the thoracic cavity and prevents the lungs from collapsing. These three autonomic MDR responses prove effective enough to enable conscious or unconscious air-breathing mammals to survive longer underwater without air than they can survive topside without air.

All whales, dolphins, and seals routinely undergo bradycardia, peripheral vasoconstriction, and blood shift during deep dives. But past a certain point even marine mammals will begin to experience hypercarbia — the rapidly rising levels of carbon dioxide in the blood. In the human drowning victim, hypercarbia eventually triggers either unconsciousness or the breath-hold breakpoint. In either case the victim finally succumbs to an inhalation.

Most inhalations of water into the trachea initiate laryngospasm, whereby the larynx constricts, preventing water from entering the lungs. Sometimes laryngospasms cause the victim to swallow the water trapped in the trachea, and the presence of seawater in the stomach then triggers vomiting, which typically produces a fatal asphyxiation on the stomach contents. At other times the laryngospasm relaxes after unconsciousness, and water rapidly floods the lungs in a process known as the terminal gasp.

In a freshwater drowning, the osmotic imbalance between the salts in the water and the salts in the blood causes this inhaled water to leach from the lungs into the bloodstream, eventually usurping as much as half the body's blood volume. If the drowning occurs in seawater, the inverse osmotic imbalance draws water out of the bloodstream into the lungs, producing pulmonary edema — in effect, drowning the victim in his or her own body fluids. In contrast to these wet drownings, some victims never make the terminal gasp and die what is known as a dry drowning.

As far as we know, dolphins — whose respiration is believed to be entirely under their conscious control — always suffer dry drownings. Consequently, the ultimate cause of their death is probably cardiac arrest due to hypoxia (oxygen deprivation).

Those who have experienced near-drowning, or have drowned and been resuscitated, report feelings of both panic and peacefulness in the course of the same incident, with peacefulness following on the exhausted heels of panic. Years ago, during a canoe accident in whitewater on a northern New England river, I came close to drowning, although I never lost consciousness and I never inhaled water. But I did spend a long period of time underwater pinned beneath a fallen log, where the force of the current was too strong for me to save myself. Rather than feeling panicked, I experienced bewilderment, followed by resignation, accompanied by the simple thought: *So this is it.* I was surprised by my own sense of calm and lack of fear, and overall the event echoed a kind of familiarity, as if I had done all this before.

According to the Tibetan tradition of death and dying, all living beings, regardless of species, and regardless of the manner of death, undergo the same multistep process. Western tradition typically considers the moment of death to occur when breathing and the pulses cease. But the Tibetans describe an internal process that continues for roughly twenty minutes beyond this — the length of time it takes to eat a meal, they say. During this period, which they call inner respiration, the dying person or animal passes through four increasingly subtle levels of consciousness.

As described by Sogyal Rinpoche in *The Tibetan Book of Living and Dying,* the first stage follows immediately upon what Westerners consider physical death. Awareness crystallizes after the physical struggle, as the dying experience a whiteness like moonlight, and "all the thought states resulting from anger, thirty-three of them in all, come to an end." The second stage appears as a redness like the sun in the sky, accompanied by a powerful sensation of bliss as all the forty states of mind resulting from desire cease to function. When this red sun and this white moonlight meet in the heart, a state of blackness, "like an empty sky shrouded in utter darkness," prevails; the inner experience is a mind free of thoughts, as the seven conditions of ignorance and delusion cease. In the fourth and final stage of inner respiration, we become slightly conscious again, says Sogyal Rinpoche, as "the Ground Luminosity dawns, like an immaculate sky, free of clouds, fog, or mist." This condition is sometimes called the mind of the clear light of death.

Here in Avamotu Bay, those spinners in the process of dying are undergoing the process of outer dissolution as their body functions fail. According to the Tibetans, their earth energies are draining away, and their minds are shifting from agitation to drowsiness. The water energies take over to support consciousness. As they too begin to fail, dehydration sets in, accompanied by feelings of irritability and nervousness, and a sensation of drowning in an ocean or a fast-moving river. Fire energies falter in the aftermath of water, as the extremities grow cold, as the mind swings between clarity and confusion, as nothing can be eaten or drunk anymore, and as no one is recognized. Fire finally cedes to the energies of the air, which support consciousness as breathing becomes labored, and as the mind trades the visions of the outer world for the visions of another world, among the crisp, white sheets of the sea.

36
Mother Ocean

BEYOND THE CONFINES of Moʻorea's barrier reef, the open ocean seems monotonously blue and empty by the standards of the coral world. From the surface you see nothing more than shafts of silvery sunlight converging a hundred feet below. Many divers shun this world because of its seeming emptiness. Most nondivers never even have the chance to see it.

Science divides the world ocean horizontally and vertically. Encompassing all the open waters away from the coasts and above the sea floor are the waters known as the pelagic zone. The nearshore pelagic is called the neritic (after Nereus, son of the Greek sea god, Pontus, and the earth goddess, Gaia), and this is what we see when we gaze from shore in most of the world. The offshore pelagic is called the oceanic zone. Lying beyond the two-hundred-meter contour marking the continental shelves, the oceanic zone encompasses not only the open ocean, but also the deep open ocean. It is visible only from a ship, or from a jet, or from space, or from the shores of mid-ocean islands.

On Moʻorea, the oceanic zone makes rare landfall. Arriving suddenly on volcanic slopes, it ascends the spur-and-groove zone, crashes onto the barrier reef, rolls across the algal ridge, and spills into the lagoon. For denizens of the coral world, the oceanic zone is their home, albeit a shallow membrane of it. Most reef life spend their adulthood bathed in its waters, while anchored or tethered to the skirts of the land. Most use the deeper oceanic zone beyond the reef as a nursery ground — a veritable Mother Ocean, where their young go for nurturing in a watery cradle.

From a scientific perspective, why exactly the larvae of corals, fishes, worms, sea slugs, sponges, crustaceans, cephalopods, bryozoans, sea stars, sea urchins, sea cucumbers, and sea squirts leave the shelter of the coral reef for the immensity of the open water is still not understood. Safety certainly is not a prerequisite, since mortality for larval reef fish in the open sea approaches 100 percent (to compensate, egg production can be many millions per adult per year). Whatever the reason, the open ocean is where most eggs go, where they hatch into larvae, and where they then join the legions of primary consumers, the zooplankton, on their peregrinations.

If you dive in the oceanic zone — or rather, if, like most everything else here, you stop and drift — you can see some of these tiny beings. By hanging under the surface, remaining still, then refocusing your eyes to a distance just beyond the tip of your nose, a world of creatures emerges — translucent things with peculiar winglike fins that refract light into tiny rainbows. Looking closer still, you may note beings so small they are indiscernible individually, appearing collectively as a kind of blur in the water.

Vision is the art of seeing things invisible, wrote Jonathan Swift, and some of these coral-reef infants afloat in the pelagic ocean are so transparent as to remain nearly invisible even in a dish of water under a dissecting microscope. Yet these tiny beings sail the currents of the globe, pursuing and capturing prey, countless trillions of them drifting in geometric precision. As you observe, some collide with you — though there's no sensation of touch in this realm where gravity and substance approach a vanishing point.

For all the tiny sailors you can see, there are many more you can't. The fields of the ocean are rich beyond measure, densely populated not only with phytoplankton and zooplankton, but also with the even smaller heterotrophic bacteria that manufacture and consume carbohydrates and carbon dioxide on a scale science can't yet imagine. The combined weight of all these bacterioplankton exceeds that of all the fish in the world ocean. One species, *Pelagibacter ubique,* possesses the most stripped-down and efficient genome known on our planet and boasts an estimated global population of twenty billion billion billion microbes. Preying genetically upon these bacterioplankton are tinier things still, recently discovered viruses known as bacteriophages, which prey on bacteria and collectively make up the virioplankton. This community reaches astronomical population densities of three billion microbes per ounce of seawater. Taken altogether, the discovery of these ultraminiatures is akin to finding the elusive dark matter of the universe.

Extending ever smaller, deeper, broader, and richer, the pelagic zone begins to embrace the properties of infinity. As you pause here, watching the largest of the small sail by, you might remind yourself that little of what is bathing you and buoying you is actually water; it's almost all life.

In the late afternoon, we dive Hauru Point, outside Mo'orea's barrier reef, on the extreme northwest tip of the island. Tiger sharks visit here on occasion, including an enormous female who appears from the deep like a zeppelin coming in to dock. At twelve or fourteen feet long, she is frighteningly enormous, with a girth that inspires awe even in our jaded French dive master.

He is a young man with an attitude of tired resignation, as if his twenty-something years on this earth, the last couple of which have been spent diving the crystalline waters of the South Pacific daily, have been debilitatingly boring. So extreme is his weariness that he cannot lift the hand-rolled cigarette from his mouth and must leave it dangling from his lower lip, trailing smoke into his bloodshot eyes. His character is so convincingly clichéd that we fondly imagine him to be our very own Charlie Allnut, Humphrey Bogart's character in *The African Queen.*

Only the appearance of this enormous tiger shark, whom he

calls Belle, brings joy to the heart of our young Allnut — as if his lover has arrived from afar for an hourlong tryst amid the pillows of *Montipora* corals. Belle is as big around as a rhinoceros, and she is so daunting that all of us experienced shark divers find ourselves flattening out on the bottom, flounderlike. But not Allnut. He rises to meet her in the blue water, wriggling upward like a tiny cleaner wrasse soliciting a nibble from a passing sea monster. In all my years making nature documentaries, I have worked with many truly brave people, but the French alone manage to produce a few pathologically heroic specimens who deeply frighten the rest of us.

Belle is nonchalant, sauntering across the blue sky of the sea, parting schools of silver-and-brown humpback snapper gathering this afternoon on the point, where the currents whiplash offshore. She ignores Allnut and descends past this spawning aggregation, zeroing in on the reef and the scent of our *aahi* chum. She passes only five feet above a camera assistant lying on the reef slope. Alarmed, he flips onto his back, holding his arms out as if to push away the passing zeppelin. It's a pathetically futile response, and we watch this small drama for which he will be mercilessly teased forever, grateful that it did not befall any one of us, each of whom would likely have responded the same way.

Except Allnut. He feathers his fins after Belle — for what purpose, exactly, none of us can imagine. There comes a point when witnessing foolish behavior at which you are grateful, albeit ruthlessly grateful, to have a camera to record the likely tragic outcome. You do not wish for the tragic outcome, though you might have a sense of its natural equity in the larger scale of things. Allnut pursues his gigantic beauty, who drifts back and forth above the tethered chum, dissecting the nuances of scent, trying to convince herself that this molecule of tuna flesh awaiting her on the reef is unlikely to harm her. Allnut hovers above, a tadpole wearing tuna scent.

Suddenly the humpback snappers, who have been gathering around an enormous gorgonian fan coral in the growing darkness, explode upward — pair after pair shooting above the reef, loosing their spawn, careening back down. The whole moving mass appears to bounce across the gorgonian in stretchy silver lines, flashing their guanine mirrors at the apex of their ascents, then retracting, like fireworks on bungie cords. It's such an incredible sight that it redirects our attention from Allnut's supersized shark.

As the pyrotechnics continue, as the spawning group replenishes itself from an endless conveyer belt of humpback snappers flowing in from the four quadrants of the compass, we are provided a feast for our eyes. Yet we miss what may be the best part of this show. Deafened by our own scuba exhaust, we cannot hear the disturbances in the flow field from hundreds of beating fins, or the crashing back to the reef inside roaring vortices of water.

But Belle, endowed with the investigative senses of a long-distance hunter, has heard too much — at least too much in such close proximity to a suspicious tuna, a few two-legged flounders, and a hovering tadpole — and she bolts down the reef slope on the propeller of her tail, fading into the particulate of depth and darkness. She has not taken the chum with her, nor Allnut, her potential chum.

Afterward, in a waterfront bar on the edge of the lagoon, where we take him for a consolatory drink, Allnut is so melancholy at Belle's abbreviated visit that our intention to commiserate soon cedes to a desire to bully. Obviously, he is not snack-worthy, we tease. But he doesn't understand the English nor any of the various French translations we offer. You must have scared her, we taunt, offering yourself like that; sharks aren't used to sacrificial victims. When we begin to suspect that he is only feigning incomprehension, we suggest that Belle does not love him, a claim that raises a private, crooked smile, his eyes squinting from smoke, the cigarette on his lip twitching.

For the fertilized eggs of the humpback snappers and all the other broadcast spawners who loose their seed into the current, the pelagic realm is a phenomenal dispersal ground. The eggs of most coral reef fish hatch a day after fertilization, and the hatchlings then spend the next seven to four hundred forty-eight days, depending on their species, sailing far and wide, grazing the pastures, and hunting the herds of the open sea.

Virtually all the fauna on the reef, from the invertebrate corals to the vertebrate fishes, produce eggs that hatch into larvae (Latin *larva:* ghost), which then disperse on pelagic currents. Even coral reef fish that brood their eggs in their mouths, or in nests on the reef, will loose their larvae, upon hatching, into the open water. In fact, of all the teleost fish inhabiting the world's coral reefs — a species count numbering, at bare minimum, four thousand — we know of fewer than ten species that do not spend part of their lives as larvae in the deep blue.

There is no equivalent in the mammalian life cycle to the larval stages of the sea, not even among our embryos, with their strange ontogeny. Larvae hatch from fertilized eggs as fully independent beings with a physical body, a habitat preference, a menu preference, and every other lifestyle aspect completely different from their adult guise. There is no period of dependency or ineffectiveness. If a larva survives the many perils of larvaehood, it will eventually undergo a profound metamorphosis in order to assume its juvenile, or in some cases its adult, form. Until then, it will raft the pelagic zone as one of the countless members of the zooplankton.

In fact, at any given time, exponentially more individuals of coral reef species are afloat in the pelagic zone than on all the coral reefs of the world combined. Consequently, some researchers suggest that coral reefs are really little more than maturation grounds for this larger open-water zone. Or, alternately, that reef fishes and invertebrates could just as accurately be characterized as planktonic animals with a relatively immobile adult phase.

Corals are also broadcast spawners, using the currents to disperse their larval offspring, known as planulae. Once a year, cued to a split-second synchronization we do not fully understand, more than one-third of the four hundred species of corals on Australia's Great Barrier Reef spawn, releasing their eggs and sperm to float upward in the water column like a reverse-flowing pink-and-orange blizzard. Most of these corals are hermaphrodites, producing both sperm and eggs in neat little packages known as gamete bundles. At the moment of spawning, the bundles rise from the corals' mouths to float to the surface on the buoyancy of the fatty eggs. Shortly afterward, the sperm detach from the bundle and swim in search of another coral's eggs.

If you fly over the Great Barrier Reef or the reefs of Papua New Guinea, Palau, Fiji, Okinawa, or the Philippines in the immediate aftermath of a mass spawning, you will get a sense of the magnitude of the event. For a few hours after dawn the sea remains a watermelon-pink, studded with silvery black seeds marking the pockets of sea surface where masses of eggs have failed fertilization and are already dying, or where they have successfully fertilized and are transforming to larval forms. As the hours go by, the colors fade until the whole looks like nothing so much as an oil slick, which it is sometimes mistakenly reported to be.

By dawn, the fertilized eggs have begun the massive developmental shift to planulae, wherein they join the ranks of zooplankton. But, because coral planulae lack developed tentacles or mouths, their time adrift is limited by their metabolic reserves. Some planulae spend a lengthy time in the pelagic zone, one hundred days or more, thanks to the endosymbiotic zooxanthellae (plant partners) bequeathed to them by their parents, which feed the planulae through photosynthesis. Other species increase their drift time by partnering with free-floating zooxanthellae soon after fertilization. A few species are endowed with the equivalent of a yolk sac.

But most larval corals are believed to spend only a few days or perhaps a week in the pelagic zone, though no one really knows for sure. A tiny few manage by sheer luck to avoid the dangers adrift: the ultraviolet radiation under the ozone holes, the surface pollutants, the predators. Floating and swimming on their cilia, buoyed by fatty lipids, and sightseeing with their chemoreceptors (taste and smell senses), these tiny survivors sail the seas awaiting the developmental changes that will enable them to recruit — in other words, to settle to the bottom and begin the juvenile phase of their lives. But they can't settle just anywhere. Settlement will mark their final home, and a Goldilocks-list of variables must be met before a planula will abandon its wandering ways.

Included in these criteria are: the type of the bottom, the motion of the water, a salinity level somewhere above 32 percent, enough sunlight for the zooxanthellae to photosynthesize, a world mostly clear of sedimentation, and a neighborhood already endowed with the presence of certain marine algae, diatoms, and bacteria. Once these criteria are met, the planulae whip their cilia into overdrive and dive from the surface to the bottom. If, upon closer inspection, the substrate is no good, they reject it and swim upward again. If it proves desirable, they attach and begin to build the little carbonate calices (Latin *calix:* cup) where they will live out the rest of their lives.

Once settled as adult corals, their task is to feed themselves — generally on all the other larvae and planulae sweeping by on the currents. With the energy from those meals, they will build the coral reef, which will, in time, provide more spawners, more gametes, more larvae, and more planulae to scatter into the pelagic zone to seed new reefs, or the same reef again.

37

Fish Tamer

I N T H E I R Y O U T H, sailing the bottomless blue of the
pelagic zone, the larval fishes of the coral reef take on
extraordinary forms. Most are tiny, as little as one one-
hundredth of an inch, and largely transparent. Many have long
decorative spines trailing above and below, no fins, and gigantic
eyes that dwarf underdeveloped bodies. They appear whimsical,
as if designed by children with crayon palettes who color in only
part of their designs before moving on to the next creation.

Until recently, most larval fish were believed to epitomize
drifting or, at best, weakly swimming plankters. But recent research shows that this is not the case. The work of Jeffrey M. Leis
of the Australian Museum in Sydney suggests that as the larvae
of coral reef fishes approach metamorphosis, they are capable of
active swimming at speeds of more than twenty body-lengths per
second, and they can hear and smell a coral reef more than six-
tenths of a mile away, move toward it, inspect it, and either accept or reject it. They are not, apparently, swept helplessly anywhere, but rather dynamically seek that which their transmuting
hearts desire.

I am acquainting myself with some of these headstrong larval miniatures — not in the open sea where they are impossible
for me to identify, but in an air-conditioned computer room at
CRIOBE (Centre de Recherches Insulaires et Observatoire de

l'Environnement), the French biological research station at the base of Opunohu Bay on Mo'orea. A photographic library of larval fish is loaded into marine biologist Vincent Dufour's computer. One image in particular interests me — the juvenile form of *Ostracion cubicus,* the yellow boxfish. In adulthood this animal will have, as its name implies, a box-shaped body enclosed in fused, bony plates. But here in its youth it looks like a tiny yellow gumdrop, gaily polka-dotted in blue, with huge yellow-and-blue eyes and small transparent fins. How can such a morsel of life get around, I wonder, how can it, as Vincent explains, navigate across ocean currents to return to an island — possibly the same island where it was spawned?

Vincent's broad brow and steady blue eyes award him a guise of earnest transparency. In the deep ocean, he says, drifting and swimming, hunting and avoiding hunters, the tiny yellow boxfish and all the others like it grew, changed, and, with the single-mindedness of salmon, returned to a home on the coral reef they had never seen before.

In fact, at this moment, they are swarming in the waters outside Mo'orea's barrier reef. Tiny boxfish, surgeonfish, *poisson Picasso, poisson dragon, i'ihi, nohu, pataitai, Cephalopholis argus, Centropyge flavissimus* — all the fish named in all the different languages. They mass by species, sorting out their own identities, even though most are as small and invisible in the water as contact lenses.

Above them, huge swells from the Antarctic roll toward the islands. Below, the stony corals of the outer reef slope hunker down to withstand the surge. Ahead lies the barrier reef, the natural buttress between sea and lagoon, and at its summit the most inhospitable zone of all, the impossible place where combers repetitively curl, break, then retreat — the algal ridge. It is this seemingly insurmountable obstacle the baby fish must breach to enter the sanctuary of Mo'orea's inner lagoon.

The young fish are anticipating the moonless night to come. They wait impatiently, as restless and skittery as flocks of birds preparing to migrate. Unlike the twilight evening of their conception, they will not travel through one of Mo'orea's passes, since too many predators are assembled there. Instead, as Vincent's doctoral research uncovered, they choose what seems a suicidal option. After nightfall, despite having never done anything of the kind before, they will gather under the humped back of a wave, ride up its powerful rise and down its thunderous landfall and surf across the algal ridge.

It's a trip that would likely kill you or me. Even today, snorkeling in Irihonu Pass on a relatively calm afternoon, I can see the power of these waves. From underwater they lift then curl, forming mercury-bright cylinders that tumble over themselves up the barrier reef toward the algal ridge. Some of the smaller waves barrel-roll all the way across, then spend their energy in the suddenly deeper water of the lagoon. But the larger swells rear and corkscrew, appearing from underwater to pull themselves inside-out until they've drawn their full height, at which point they hurtle onto the ridge, exploding in billows of blinding white bubbles. It's difficult to imagine the tiny fish, delicate as flower petals, surviving such a journey. Yet they do, and once inside the lagoon, these youngsters swim in search of reef patches on which to settle. Since the best sites are already commanded by adult fish, some newcomers choose less desirable or less developed areas.

I've discovered one such place alongside a crumbling rock jetty inside Mo'orea's lagoon. Apparently there is too much boat traffic here for the larger fish, and also too many human fishers. But for the new arrivals it's perfect — the disintegrating cement of the pier porous with hidey-holes, the bottom strewn with old tires, PVC pipes, soft drink bottles, shoes. This dump turns out to be a wonderland for the babies of the coral reef, and in the coming days I snorkel here over and over again, never quite able to get my fill.

In fact, everything you could ever hope to see on the big reef is here at a fraction of its normal size. A green-spotted moray eel no bigger than my pinkie snakes out of a chip in the dock. A squadron of spotted eagle rays, each smaller than my outstretched hand, cruises the wall of the pier, their striking designs of white circles on black backs flexing with each wing stroke. Inside an eight-ounce can that once held *jus d'ananas* (and exactly mimicking its colors), lives a tiny yellow-and-white octopus, pale and translucent as a ghost.

All the fish are here, too, most no bigger than dimes, but already, within just a few days of settling on the reef, they have blossomed into their adult color patterns. Over there, by a young colony of pink *Pocillopora* coral drifts a *huehue*, the tiny pufferfish whose radiating lines of green-and-orange eye makeup give it a look of happy surprise. Over here, foraging in the red sponge growing out of a plastic sandal, a miniature bird wrasse the size of a dime, whose long beak is probing in search of miniature crustaceans.

Occasionally an adult of something or other swims through, amplifying the illusion that I've arrived in a watery Lilliput. A six-foot moray eel startles me with its unexpected daytime excursion from its home cave. I find myself diving to the bottom for cover, then laughing at such an overreaction. The fish, however, have no doubts. All the bright clouds have vanished into the honeycomb of junk around the dock.

Vincent Dufour's most remarkable photograph is of his outstretched hands cupping a wriggling treasure of about fifty of these jewellike young fish. They are all instantly recognizable in their adult patterns: butterflyfish, triggerfish, angelfish, pufferfish, boxfish, surgeonfish, filefish.

This photo, displayed in Vincent's press kit, describes the premise behind his new venture, which marks his own personal migration from science to commerce. He has become a businessman, he says, and has returned to Mo'orea from his home in France to launch an enterprise called Aqua Fish Technology, which he believes will make money, but beyond that, which may eventually help save coral reefs from some of the many assaults of the modern world.

We stroll across the grounds of CRIOBE on a hot, sticky

southern-hemisphere summer day, the air around us infested with the hordes of mosquitoes that the research station inadvertently breeds in the same aquaculture ponds where it rears brine shrimp. Black thunderheads tangle in the jagged peak of Mount Tohiea, charging the air. Since Vincent's return, he has initiated a small-scale building boom, transforming one of CRIOBE's old sheds into a series of rooms filled with racks for aquariums of various sizes, as well as sinks, sorting tables, pumps, and filters. Upstairs, two tiny bedrooms house his staff, the young men and women eagerly awaiting the chance to nanny the baby fish around the clock. Outside, the groundwork is laid for two much larger tanks, and nearby, the ground is littered with pieces of Vincent's own creation, the keystone to his whole enterprise, a tubular steel contraption he calls a crest net. Mounted atop the barrier reef, this net will collect virtually all the little fish that surf across at that point.

Since there is no immediate hope of breeding reef fish in captivity (because, at the moment, there is no way to recreate the conditions of a *mascaret* flowing through the pass in a barrier reef), Vincent will leapfrog that obstacle. He will capture the little fish as they return to the reef and raise them in the thirty-five hundred cubic feet of aquariums under construction. Because he is one of perhaps only ten people on earth who can identify these postlarval fish species, he will be able to sort and separate them into quarantine tanks, where the carnivorous species will not consume the herbivorous ones. Tailoring their food supply, he will domesticate them until they are well-behaved enough to introduce into home aquariums, without fear of them nibbling and destroying the hobbyist's expensive, decorative corals.

38

A Force Like a Hundred Thousand Wedges

TACKLING THE WORLDWIDE trade in aquarium fish might not seem a particularly profitable or even noble task, yet it is an industry desperately in need of transformation. Currently worth at least five hundred million dollars a year, with rare species selling as high as twenty-five hundred dollars per fish, this trade is fueled by some two million home aquarists, half of whom live in the United States, one-quarter in Europe, and the rest scattered everywhere else on the globe. As many as twenty million fish are traded each year, representing as many as one thousand species, and this growing market proves rabidly destructive to the coral reefs of the world's poorest nations.

Many aquarium fish currently arrive in the developed world after surviving the most brutal collection methods. Subsistence fishers throughout the tropics destroy entire ecosystems in order to capture the few stunned fish surviving on the perimeters. One of the primary collection methods for the aquarium trade is stupefactant poisons, mostly sodium cyanide, but also bleach and detergents. Sodium cyanide begins to kill corals within thirty seconds of contact, and the remaining marine life succumbs soon after. A handful of fish at the outermost edge of the destruction — disabled but not dead — are then collected by hand. In desperation, as fish stocks dwindle, some fishers dump entire fifty-five-gallon drums of cyanide onto reefs, hoping to reap the few fish remaining, or the few that have returned.

With each purchase of a beautiful lionfish, or a plucky butterflyfish, the home aquarist, obliviously or uncaringly, funds the devastation. Yet if Vincent Dufour's plan succeeds, he will be able to offer an alternative to these practices, with a low-impact and relatively cheap technology easily installable along the impoverished coastlines of the tropical world.

I question him about something bothering me, though, the impact of removing the newly arriving fish from Mo'orea's lagoon. He says that 90 percent of these youngsters are destined to die within a few days of surfing and settling. Still, I wonder, aren't they destined to die inside the stomach of some larger, hungry reef fish who depends upon them for survival? He admits his enterprise may not be up to the standards of deep ecology, but his crest nets will be mounted on only a small portion of the perimeter of the barrier reef. Many more fish will surf across without getting caught.

Beyond the aquarium trade, he hopes eventually to wrestle with the problem of the live reef food fish trade (LRFFT). In this, fish are generally collected with explosives rather than poisons. Fishers toss beer-bottle bombs filled with dynamite or homemade fertilizer/kerosene mixes onto the reefs. The resulting explosions kill fish by destroying their swim bladders, enabling human fishers to collect those along the perimeters who are only disabled. In addition to killing countless fish, the blasts also reduce the reefs to rubble, from which they will likely never recover. Some reefs in Southeast Asia are currently sustaining multiple blasts per hour.

The LRFFT is fueled largely by an Asian dining preference for fresh fish, particularly fish from afar. The species most in demand are various grouper and the large napoléon wrasse, which are flown alive primarily to Hong Kong, but also to Taiwan and China. Procured by pricey restaurants, these captives swim in dining room aquaria, awaiting the moment when a customer will point one out, whereby it will be fished from the tank and steamed alive.

Hong Kong imported thirty-two thousand tons of live fish in 1997 at wholesale prices of up to eighty-seven dollars per pound, for an annual total of five hundred million dollars. Prices at the retail level run even higher, with live fish costing 800 percent more than frozen fish. The favorite fish are those, perversely, known to be endangered species, which fetch even more money. The larger the fish, the greater the perceived status of the buyer, who may be trying to impress his or her guests at a wedding banquet or a business banquet. Seven-foot-long live napoléons have reportedly sold for more than ten thousand dollars, while the supreme delicacy, a plate of nothing more than the thick lips of the napoléon, sells for as much as two hundred twenty-five dollars.

As more people wish to demonstrate their wealth by ordering extravagant food in public, the hunt for declining species expands to all the world's reefs, making the LRFFT the greatest immediate threat to grouper and napoléon populations across the tropics. Demand grows in inverse relationship to species' declines, with many fish now targeted during their spawning aggregations, wiping out entire adult populations and all their potential progeny. Some species have been locally extinguished in the course of only one or two spawning events.

In addition to the destructiveness of the poison fisheries and the blast fisheries, coral reefs also face ruination from the *muro-ami* drive fisheries, which overlap poison and blast fisheries in many places. The *muro-amis* employ groups of dozens or hundreds of breath-hold divers — usually children working in conditions of slavery — to swim beneath large nets suspended at the surface. Using rocks hanging from lines or hand-held poles, these throngs of wriggling children, equipped with little more than swim goggles, smash the living reef to pieces, forcing panicked sea life to flee into the waiting net.

Not only is the mortality rate for sea life unsustainable in the *muro-ami* fisheries, but many children drown in the process, and although illegal worldwide, these fisheries persist. In some places, they have been replaced by *paaling* fisheries, where free divers, holding compressed-air hoses in their hands, drive reef fish into waiting nets ahead of curtains of bubbles. This practice is less damaging to corals but still unsustainable for fish life.

Currently, at least forty countries withstand blast fisheries on their coral reefs, more than fifteen nations suffer cyanide fishing, and the numbers are growing. Meanwhile the *muro-ami* and *paaling* fisheries, once widespread only in Southeast Asia, are now making inroads along the poor coastlines of Africa and the Caribbean.

The effects of poison fisheries, blast fisheries, and drive fisheries are known as Malthusian overfishing, which occurs when the number of human fishers overwhelms the sustainability of the fishery, yet fishing continues anyway, in ever more destructive and desperate ways, until the reef is destroyed. The dubious naming honor goes to the English political economist Thomas Robert Malthus, who predicted that world population would exceed world food supply, based on the fact that population increases geometrically, whereas food supplies grow arithmetically.

Charles Darwin credited the final consolidation of the scattered pieces of his origin of species to Malthus's *An Essay on the Principle of Population: A View of Its Past and Present Effects on Human Happiness; with an Inquiry into Our Prospects Respecting the Future Removal or Mitigation of the Evils which It Occasions*, published in 1798. In a passage from Darwin's personal notebook, written while reading Malthus, his excitement is obvious. "One may say there is a force like a hundred thousand wedges trying [to] force every kind of adapted structure into the gaps in the economy of nature, or rather forming gaps by thrusting out weaker ones."

The majority of the world's coral reefs fringe the poorest nations on earth, where poverty is exacerbated by growing human populations. Forced to make a living any way possible, the human fishers of these coastlines have, understandingly, become one of the hundred thousand wedges. Their use of some or all forms of Malthusian overfishing enables them to collect their neighborhood's last fish, which are then sold to wealthy nations with expensive dining habits and extravagant hobbies.

As the human population swells toward nine billion by 2050 (2.6 billion more than 2005's population, and an increase larger than the entire world population in 1950), Malthus's concerns resound. His primary motivation was to "investigate the causes that have hitherto impeded the progress of mankind towards happiness." Chief among the impediments, in Malthus's view, was the "unequal distribution of the bounties of nature, which it has been the unceasing object of the enlightened philanthropist in all ages to correct."

More than two centuries later, as the bounties of nature grow ever more unequal, the poor people of the coral-reef world are forced to sell the source of their own sustenance and their own happiness to people faraway. The wealthy buyers, meanwhile, are hoping to purchase happiness for themselves, as measured in the envious eyes of their neighbors. And so the wild fish of the tropical seas — the objects of everyone's desires — are removed from the coral reefs they enliven and define, becoming the first link in a growing chain of unhappiness girdling the globe.

Vincent Dufour believes that the fish species prized in the LRFFT can be captured as larvae, raised in captivity, and exported to Asian markets worldwide, thereby alleviating some of the pressures on living reefs and wild fish, and perhaps making the buyers in the chain of unhappiness happier. He also hopes to raise fish for local consumption, perhaps making the suppliers in the chain of unhappiness happier.

Because even in French Polynesia, where human population pressures are less extreme than elsewhere in the coral-reef world, and where the people enjoy a measure of prosperity unknown in countries like the Philippines, most of the heavily populated islands, including Mo'orea, are overfished. In fact, most of the Society Islands have come to depend on the relatively pristine fish stocks of the outer atolls of the Tuamotus to feed their own residents and tourists. Nowadays, many interisland commercial flights within French Polynesia stop at remote atolls for the sole purpose of transporting fish. The scene is always the same. The stop is rarely announced in advance, and then suddenly your plane is descending toward an unfamiliar atoll, where the locals wait beside the landing strip in skiffs. All the passengers disembark to stretch their legs, as cooler after cooler of reef fish are loaded aboard, bound for the restaurants and hotels of Tahiti, Mo'orea, and Bora Bora.

In addition to raising larval fish to adulthood and making them available for local consumption, Vincent Dufour believes his captive-raised fish can alleviate the serious and growing health threat of ciguatera poisoning. Wild fish are infected by eating a poisonous dinoflagellate, *Gambierdiscus toxicus,* which flourishes on damaged reefs where dying corals are being replaced by algal turf. Vincent will feed his larvae food specially prepared to be free of *Gambierdiscus toxicus,* thereby guaranteeing local consumers a safe food supply.

In Mo'orea, where I snorkel each morning off the beach at the Village Faimano, the future for wild fish looks ominous. Algae bloom so rampantly that in many places the fringing reef inside the lagoon is akin to a city on the verge of collapse. The skeletons of huge plate corals, killed in the cyclones spawned by El Niños, lie broken and tilted, or toppled upside-down, as if chandeliers had fallen from on high. Thick mats of yellowy fibrous algal turf, ropy as old cobwebs, coat the debris. In between, vast stretches of barren sand prettily decorated with nothing more than dancing hexagonals of sunlight creep up to and over the petrified remains of dead cowry shells, a moray eel strangled in fishing line, or the empty, gaping maws of giant clams.

Lovely things still reside here, too, dense schools of herbivores — convict surgeonfish, cerulean damselfish, goldtail demoiselles, humbug dascyllus. Occasionally pairs of ornate butterflyfish, breathtaking in orange and yellow, flit by. A Picasso triggerfish, as bold and abstract as its namesake, chases me away from its nest in the sand. But here, in microcosm, is much that is wrong with the world's reefs.

Yannick Chancerelle, a staff biologist at the French research station CRIOBE, is monitoring the health of coral reefs on the islands of French Polynesia. He is a young man with a melancholy demeanor. Perhaps living in this beautiful place and chronicling its demise, he finds his happiness slipping away with it. Yes, he agrees, Mo'orea is a good example of all the things that can go wrong when reefs and humans come together. To begin with, the island is becoming a bedroom community for the nearby island of Tahiti and the territory's capital, Pape'ete. Fed up with traffic on the colonial-era road system, people are flocking to Mo'orea, where they build houses and commute back to work in Tahiti aboard fast ferries that make the trip in twenty minutes.

The growing population contributes to a problem nearly ubiquitous in the tropics. The island of Mo'orea has no sewage treatment, depending instead on septic tanks. Sooner or later, these tanks leach human waste into the lagoon, where the nutrient-loading acts as a powerful fertilizing agent, energizing the growth of the algal mat. As the turf grows, it smothers the living corals, which in turn fosters the ciguatera explosion.

Even on Mo'orea's outer reef slopes, where strong oceanic currents constantly refresh the corals, there is ample evidence of the human impact. The silt from poorly planted pineapple plantations works its way through the passes, burying parts of the reef slopes and their ancient corals. Moreover, the repeated anchor drops of fishing boats smash old coral heads and leave their remains scattered on the ocean floor like the shards of a once-great civilization.

Even scuba diving, considered a form of ecotourism and therefore good for wild places, has a subtle downside. The impact from its rapid growth is greater than just the physical damage caused by yet more anchor drops, and by the careless divers who collide with and clamber over the living coral. The very nature of the sport is changing, with many dive trips evolving into ersatz "extreme" shark-diving trips. Many dive masters now carry bait boxes with them at all times, and the marine life is so well trained that before the divers even hit the water the sharks are waiting — scores of blacktip reef sharks, gray reef sharks, the occasional lemon, and now and again, frightening everyone, a huge tiger.

Shark trips are even offered inside Mo'orea's lagoon, where nervous snorkelers, some wearing lifejackets in water only four feet deep, board outrigger canoes for the short trip out to where baby blacktip reef sharks flock to the offerings of offal. Stingrays have caught on, too, dozens of them pursuing the free food by slithering up the bodies of snorkelers, who emit the most astonishing range of not-quite-human vocalizations. All of this is fun, I suppose. But I worry that we're trading an ephemeral rush for the reef's real wonder — the rapture and stillness that comes from being in the presence of this realm's deep and protective mysteries.

39

Gleanings

J UST AS WITH the wonders of the reef, most of its troubles are also beyond our direct sensory experience. Hovering above an enormous yellow anemone, marveling at the way its resident colony of clownfish dart in and out of the dancing arms of its poisonous tentacles — can you judge the effects of the Jet Ski that just tore by overhead, of the fuel that leaks from its two-stroke engine, of its noise pollution on marine life that depends on sound for offense or defense, of the inevitable collisions with coral heads at low tide, and of the surface fish left like roadkill in its wake? Or, for instance, how really to understand the outcome of decisions made half a world away in France that award citizens there tax breaks for investing in hotels or businesses for which there may be no need in French Polynesia, or else no means of support, such as an adequate water supply?

Today we are entering an era of great threats to coral reefs. Under assault from pollution, coastal development, agricultural runoff, overpopulation, and overfishing, reefs exhibit their vulnerability in many ways. Each year new coral diseases are identified — black-band disease, white-band, yellow-band, red-band, patchy necrosis, white pox, and on and on — some caused, frighteningly enough, by the desertification of Africa, by huge volumes of dust in the atmosphere dropping bacterial and fungal spores into the weakened seas. A study conducted by Catherine Drew Harvell of Cornell University, and others, concluded that the warming climate is triggering unprecedented numbers of disease outbreaks from a full spectrum of agents, including viruses, bacteria, fungi, and parasites, in habitats ranging from coral reefs to rainforests. Coral reefs suffering bleaching episodes are particularly vulnerable to these pathogens.

Furthermore, because of their reliance on photosynthesis, reef-building corals can grow only inside the sunlit zone, in the uppermost membrane of water. But as global temperatures and global sea levels rise, that sunlit zone may eventually stretch beyond reach. The geological record of the most recent great flooding, at the close of the last Ice Age ten thousand years ago, shows that while some reefs grew fast enough to keep up with rising sea levels, others did not. Those that failed were presumably already too stressed from sedimentation, nutrient-loading, or changing temperatures — the same problems plaguing virtually all reefs today.

Meanwhile, the ultraviolet light now pouring in through the ozone holes above the planet's poles not only stresses corals enough for them to bleach, but also kills zooplankton, including the delicate fish larvae in the pelagic zone. Losing this, the true nursery ground of the sea, has profound ramifications for the entire global food web — both marine and terrestrial — and for all the people worldwide, including you and me, who utilize more than one hundred million tons of food from it yearly.

But for the people who live alongside coral reefs, the effects are devastating. Barrier reefs, as their name implies, form a stronghold between the land and the sea, protecting humans and their homes from the onslaughts of waves. Coral reefs are critical to the survival of low-lying areas, and assessments of the December 2004 tsunami found that coastlines buffered by healthy, well-developed reefs suffered less damage from the waves than coastlines with no reefs or dead reefs. But while corals help buffer the coasts from storms, they also, obviously, suffer the effects, and too many storms in rapid succession can ruin even healthy reefs.

A 1997 study struggled to estimate the total value of the world's coral reefs by factoring in four types of benefits: consumptive uses of reefs such as coral mining, fisheries, and shell fisheries; nonconsumptive uses such as tourism and recreation; indirect use values, including the way coral reefs protect coastlines and serve as fish nurseries; and option values, in other words, something you will pay some amount of money to protect now because it may yield you or your children some value in the future. The total yearly value arrived at was three hundred seventy-five billion dollars.

Whatever their true economic worth, there is no doubt that in the absence of reefs, the poor people of the tropics will become poorer. Without reefs to keep them afloat, atolls will disappear. Without reefs, many fish and shellfish — in most cases the only source of protein for people on tropical coasts — will disappear. Even if Vincent Dufour's dream comes true and aquaria for baby fish appear across impoverished tropical coastlines, without healthy reefs, without a sustainable population of adult fish, the moonless nights will come and go without spawners, until the last little fish surfs across, and then no more ever again, for him or anyone else to catch.

I'm dining with biologist Michael Poole, his Tahitian wife, Mareva, and their sons, Temoana and Tearenui, in the family home on acres of quiet land at the northeast tip of Mo'orea. Here, where the barrier reef comes to within sixty feet of their beach, they relax in the evenings and watch as courting groups of humpback whales travel by, and again in the mornings, as Michael's study animals, the spinner dolphins, return from their nighttime feeding grounds in the oceanic zone to the safety of the lagoon.

This evening we chat, drink wine, shuck shrimp, and listen to the ceaseless swoosh and thunder of waves breaking and retreating on the reef. Mareva reminisces about her childhood, how her father caught all the pretty little lagoon fish in the water just off their *fare*. These fish are hard to find now, and she and Michael discuss the fact that their sons have never even eaten some of them — little goatfish, for instance, which Mareva claims have too many bones for Michael, a Westerner, to eat safely. Michael strenuously disagrees. He grew up in hot, watery places from Florida to Bermuda and knows how to eat fish. Not these fish, says Mareva. You have to be born eating them. Michael counters that she can't eat a chicken nearly as well as he can, doesn't know how to pick *those* bones clean. They laugh, they're charming, the cultural divide, you suspect, an endless source of wonder and periodic exasperation to them.

It's difficult to comprehend the magnitude of threats facing the world's reefs, and here, over this pleasant dinner, we discuss the smaller troubles, the ones intimate to Michael and Mareva's daily life. In Mareva's girlhood, fishing was a community activity, with each family member inheriting a valued role. When he first met her family, Michael was sent out with Mareva and her sisters to the reef at low tide, each armed with a bucket and small tool to glean mollusks for the meal. Although he worked hard — clambering on the slippery surfaces, struggling to locate the shellfish, and inexpertly prying them from their holdfasts — when they returned to shore his bucket held a meager prize of only a half dozen oysters. Mareva's and her sisters' were overflowing. In this way, Michael learned the value of her skills.

The coral reef in traditional Tahitian society was a source of social stability, granting everyone a fun and rewarding job. Although many Tahitians still reef glean, the rules are changing. Mareva says that once there was an etiquette, an unspoken understanding, that the reefs nearest your own dwelling were your reefs, not to be exploited by others without an invitation. Nowadays, not only does the ferry deliver new residents, it also brings crowds of weekenders from Tahiti, the Sunday fishermen, the spearfishermen, the Jet Skiers, the gleaners, who descend upon whatever portions of the reef they desire without asking permission, thereby altering both the lifestyle and the community of the island. As people here become more migratory, they feel less attached to, and less responsible for, the fate of their natural neighborhood.

Yet many in Moʻorea are still deeply engaged with their *fenua,* and the next morning Michael stops by my *fare* to point out a small protest under way in the lagoon, a dozen or so outrigger canoes, fishing skiffs, and plastic dinghies tied together offshore, all bedecked with beach umbrellas and packed tight with Moʻoreans. They have tethered themselves to a yellow dredger, which for the past few days has been illegally pumping sand from all around the lagoon to replenish the beachfront lost in the construction of a new Outrigger Hotel. The Moʻoreans are upset. The sand being taken is their sand, undermining their own beaches and further degrading their reefs. They are determined to prevent the dredger from resuming its work, and have enlisted a contingent of supporters onshore, who hold up signs written in French and Tahitian: DON'T TOUCH OUR SAND; THINK OF OUR CHILDREN.

It's a tiny demonstration, yet as the days go by it makes the local newspapers and the television news, thanks to the tenaciousness of the protesters, and the feeling that — ensconced in their small boats, cheerily trading picnic food back and forth, slipping in and out of the water to cool off — they might, rather happily, stay afloat forever. Meanwhile, the hotel continues its noisy construction of a long line of overwater bungalows stretching far out into the lagoon, ruining the neighborhood views and disrupting the natural waterflows, so that all the beaches in the area continue to erode, with or without the dredger.

The protest seems both sad and hopeful. Sad because it's too late to succeed, as the Territorial Government invariably chooses to accept token fines in lieu of environmental compliance. Hopeful because I can feel the power in this small gathering, can feel that their resolute action might spawn something that will grow, spread, and someday come back in force.

40
Across the Threshold

I N T H E M I D D A Y H E A T, when shops and businesses close their shutters and most Moʻoreans retreat for a nap, the protesters slump under the thwarts of their dinghies, the occasional hand reaching over the side for a flick of cooling water. Even fish seek relief from the sun. Snorkeling to the middle of the lagoon, I find the underside of a diving platform crowded with pufferfish, rabbitfish, Moorish idols, lemonpeel angelfish, and a pair of teardrop butterflyfish. All are assembled in the shade, swaying to the jingling accompaniment of the platform's mooring chain, a hypnotic water chime driven by lazy waves at the surface. The fish face the current, backpedaling against it with the rotation of their pectoral fins, as if dozens of fans are silently whirring.

The only large motion in this sleepy environment is a parrotfish the Tahitians call *patti aʻa*. Dark yellowish-orange flanks make this species look sunburnt, and, as if to prevent more of this, he is also trying to get into the shade of the dive platform, but to no avail. To approach he needs to bypass a large *Acropora* coral, from which a squadron of dusky damselfish (*Stegastes nigricans*) — charcoal-colored fish many times smaller than the parrotfish — emerge and repeatedly drive him away. Earlier, the damselfish came after me, nipping at my fingers while clicking angrily at me, forcing me to swim by with my hands tucked into my armpits.

These unlikely defenders are one of the coral reef's keystone species — defined as an organism with a greater role in maintaining ecosystem function than seems obvious based on its abundance. Doctoral research conducted in the 1990s by Mary Gleason of the University of California, Berkeley, found that Mo'orea's fringing reef, which had been badly battered by the 1982–83 El Niño cyclones, was (and still is) recovering more quickly inside damselfish territories than anywhere else.

The reason is the aggressive territoriality of these little fish, who guard their homelands from anything and everything that swims or crawls by. By driving away the *patti a'a*, the damsels are effectively protecting all the young corals in their *fenua*. If not for their vigilance, the parrotfish would linger here, excavating its favorite algae from the substrate and incidentally scraping infant coral colonies from their holdfasts. Little by little, the grazing of these parrotfish and other herbivores would reduce the overall coral cover in the area.

Curiously, though, the damselfish are not protecting corals at all, only their own landholdings, the gardens of filamentous red algae they maintain on dead coral heads. By meticulously weeding unwanted plants, such as sargassum and turbinaria, from their gardens, the damsels are also controlling the algae that are the primary beneficiaries of septic tank fertilizers, and the ciguatera fish poisoning they foster.

Yet the effects of dusky damselfish on their ecosystem don't end there. Their gardens also form doormats that trap organic sediments and act as culture mediums for the growth and development of tiny invertebrates and for the bacteria that supply vital nitrogen to the damselfishes' gardens. The duskies eat the algae in their gardens, as well as some of the gastropods (snails), sponges, and copepods that thrive in the doormat. In this way, their world is circular and self-sustaining, and maintained with enough Swisslike order to earn dusky damselfish their other common name, dusky farmerfish.

All told, these hard-working horticulturalists modify the shallow-water environment in extraordinary ways, enabling the recruitment and survival of a far more complex assemblage of corals and coral reef dwellers than exists elsewhere on lagoon reefs. If not for their ceaseless efforts, other more dominant forms of algae would prevail, including coralline algae that might crowd out the reef-building corals.

And, so, visiting the coral world, you might notice a damselfish colony and perceive little more than a gathering of small, strangely pugnacious gray fish. Yet in the accumulated tasks of their daily lives, in their obsessive — and to us, nearly invisible — industry, these fish bear a critical responsibility for the regeneration of damaged reefs. Braving the damselfishes' wrath, I untie the pencil from my underwater slate and drop it into the center of their algal lawn. A buzz ripples through the colony, and the fish actually look offended. One damsel darts in, grabs the flotsam in its mouth, and ferries it away from the farm to drop it on the open sand well clear of the coral head. Then he hurries back to join the others scolding me with their clicking, and angrily facing me down.

At the end of the *Mahabharata*, India's epic poem of birth and death and birth again, the monumental Battle of Kurukshetra takes over the earth. For eighteen days, the battle rages back and forth. The culmination of the fight marks the beginning of the *Kali Yuga:* the fourth and final age of humankind, when our noble morals and ideas crumble. By the reckoning of the *Mahabharata,* the four ages of humankind comprise 4.32 million years — an eon for us, which nevertheless translates into only one one-thousandth of a single of Brahma's days.

At the end of each of his days (4.32 billion human-years), as Brahma sleeps, all the realms of the cosmos enter a state of suspended animation within his mind. The next morning, he recreates the realms exactly as they existed before the night. Thus Brahma lives for fifty years, with each year composed of three hundred sixty days and three hundred sixty nights (each spanning 4.32 billion human years), awarding him a lifespan of 155.5 trillion years.

When night falls on Mo'orea's reefs, the dusky damselfish abandon the day's work to seek shelter inside the protective arms of staghorn corals. From one point of view, these staghorns are the products of Brahma's mind. From another, they are the products of the damsels' own ceaseless efforts. Since many herbivorous fish choose to sleep inside the labyrinths of branching corals, the duskies are joined in bed, or in neighboring beds, by other damselfish, angelfish, filefish, triggerfish, butterflyfish, unicornfish, and surgeonfish.

As these herbivores nestle down to sleep, their offspring wait offshore. Some of these tiny travelers are already partway through their postlarval transition and are beginning to don their juvenile colors and patterns. Others still wear the ghostly costumes of the pelagic zone. Whatever their degree of coloration, they are poised to slip onto the reef under the cover of night. Since every inch of suitable ground is already claimed by adult or juvenile fish of their own species, they will stake their claims in darkness.

The combers rise and curl, striking the algal ridge with the force of a hundred thousand wedges. The little fish wait, skittery and untested. At the right moment, whatever that is, acting on cues we might only imagine, they surge forward at the rate of more than twenty body-lengths per second, hopping the back of a wave and surfing over the algal ridge into the shelter of the lagoon. Tasting their way forward, they follow the scent of pheromones produced by the adults of their own species, until they come upon the staghorn bed where other dusky damselfish sleep. While the adults slumber, the tiny, mostly transparent newcomers wriggle into bed beside them.

Thus the youngsters greet the adults, who awaken in the morning to discover they have spent the night with a swarm of little ones: a de facto admission of acceptance in the world of fish. In this manner, with the slick practice of evolutionary precedent, the transmuting larvae find themselves a home in a neighborhood they have never seen before, the worth of which has already been vetted by real-estate-owning adults successful at a lifestyle the youngsters have yet to actualize.

Most every night these ghost larvae, or others like them, surf across Moʻorea's barrier reef — and each night other beings infiltrate the lagoon alongside them, or else exit the lagoon through one or another of the twelve passes. As it always has, life flows over the threshold. Fish come and go, to sleep in one world or to hunt in the other. Nesting turtles come. Plankton-picking fish go. The barracudas schooling in the passes by day exit onto the outer reef slopes. The tide coaxes travelers back and forth, as does the presence or absence of the moon.

Through whatever passes they entered at dawn, the spinner dolphins exit, cranked up on the joint engine of their adrenaline. Individuals tumble through the air and through the water in their race to the deep. Young and old, male and female, they spin their way offshore, where the groups meet and mix, break apart, move off, regroup. As intensely bound as spinner society is, it is tied not only by kinship. "To a significant degree," Ken Norris writes, "they have become a society of remarkably cooperative friends."

For the spinners trapped inside Avamotu Bay, the south-southwest swell continues. Its relentless force draws a blinding curtain of bubbles across the pass, through which no dolphin can see, hear, taste, or smell. Perhaps this bubble net, which is slowly engineering their demise, also offers privacy. Perhaps they do not have to hear the cacophony of their working friends offshore.

In the coming days, one by one, the number of entrapped spinners dwindles until none is left inside the bay, and none is ever seen again outside the bay. Caught in the sieve of Moʻorea's barrier reef, they remain part of its energy, perhaps inside the belly of a lemon shark, or a scavenging lobster, perhaps until the next great churning of the ocean relocates their molecules elsewhere.

Epilogue

You may bite the sugarcane, break its joints,
Crush out its juice, and still it is sweet.

— *Naladinannurru (The Four Hundred Quatrains)*

Teti'aroa French Polynesia

THE SEVEN-TENTHS of our planet that lies underwater is effectively another world within our world, one we still barely know, familiar to us mostly at its margins and through the distortions of the surface. It does not require much imagination to understand how interconnected the many dominions of the sea are, from the benthic sea floor to the open pelagic zone to the coastal reefs. It takes perhaps a greater leap, something more akin to faith than to knowledge, to comprehend how inextricably bound these waters are to our own terrestrial existence.

We have focused much in recent years on rainforests, and coral reefs are frequently described as the rainforests of the sea — likewise recognized as genetic nurseries from which new life forms continuously well forth. Scientists estimate that a greater density of species exists on coral reefs than in rainforests, and that a greater biodiversity exists within the higher taxa on reefs than in any of the earth's other megadiversity ecosystems.

Coral reefs are powerful arbiters of life both in the sea and on the land. The oceans they help stock are the chemical engine driving the planet, stabilizing our climate, refreshing the air we breathe, making the rain that feeds the rivers and lakes, which water the crops upon which we depend. This water world, and its most fertile and fragile edge, the coral reefs, are the continuing cradle of life on earth.

Pondering these things in one of the most pristine places I know — the tiny atoll of Teti'aroa in French Polynesia — I find some comfort in the long geological history of our planet, the knowledge that life comes and goes, arises and changes on a schedule that has been, at times, extremely violent even in our absence. Reefs built by scleractinian corals (our present reef builders) chronicle the past two hundred forty-five million years of earth history, while earlier reef builders (cyanobacteria and sponges) chronicle billions of years earlier. Yet the history of these ancient reefs is not contiguous, but rather one of fits and starts, explosions of life followed by long, reefless epochs. In fact, throughout time, reefs have suffered extinctions or near extinctions about a million years in advance of the better-known mass extinctions of terrestrial life forms, including the dinosaurs — raising doubts about the meteor-impact theories.

But whatever the meaning of the reef's demise, the geologic record indicates a predilection on our planet toward the emergence of reefs, toward having something in the sea that builds habitats of such enormity and opportunity that much other life evolves around it. The greatest achievement of modern reefs is Australia's Great Barrier Reef, which is visible from space and has conceivably already informed alien intelligences that our planet harbors life.

Sitting on the sugar-white sand under the lacy shade of coconut palms, watching the life beneath the water as it appears on the surface as ripples of motion, I feel that if the geologic record teaches us anything, it shows that with or without either our help or our hindrances the earth's reefs will go and then come again, as long as the sun delivers its rays and the waters flow. Whatever role we might play in the next great extinction will surely have less effect on the tenacious reemergence of reef-builders than it will on us. Reefs, we know, can survive without us. The opposite may not be true.

Teti'aroa, in its most recent incarnation, was Marlon Brando's island, bought after the filming of *Mutiny on the Bounty*, when he fell in love with Tahiti, and with his beautiful costar from Bora Bora, Tarita Teriipaia. He first saw the atoll from the top of the island of Tahiti, twenty-six miles away, "a slender pencil of land lying on the horizon," he wrote. "Before long, it was exerting as mystical a pull on me as Tahiti itself."

Before Brando purchased it, Teti'aroa had been, at various times, exile to a leper and a copra plantation. When Brando first laid eyes on it, it was the lifelong home of Madame Duran, an elderly woman whose Canadian father had been given the island by Pomerae V, the last king of Tahiti. In the 1960s, Madame Duran lived alone on the atoll with a single woman helper. Since she had been blind for twenty-five years, evidence of her sight-seeing aids were (and are) still evident on the *motu:* the scars in the trunks of the coconut palms marked by the wires she followed from tree to tree, while delicately holding a rag in her hand.

"The happiest moments of my life have been in Tahiti," Brando wrote. "If I've ever come close to finding genuine peace, it was on my island among the Tahitians." Promising Madam Duran that he would preserve the atoll, he kept development low-key, eventually building the tiny Hôtel Teti'aroa Village, which was run by Tarita and her family until Brando's death. At any given time, one or more of his eleven acknowledged children visit or reside on Teti'aroa's only inhabited *motu*, where they fuel rumors of profligacy — including late-night orders to Pape'ete for helicopter-delivered pizza. Like the coral reefs surrounding his atoll, Brando was, in his own way, a broadcast spawner, procreating with beauties in all cardinal directions and producing children who specialize in extravagance.

Whatever spell Teti'aroa exerted on Brando it exerts on many. Above all other islands, this atoll displays a world of blues you might never expect to see in this mortal lifetime: peacock blue, baby blue, azure, sapphire, beryl, turquoise, aquamarine. Perhaps these are the colors awaiting us as heaven's first surprise, what the French call *bleu lumière*.

Sitting on the beach, I try to convince myself that I can understand the many mysteries of the coral reef nearly as well, right here, right now, as if I were underwater. Not just the mechanics, but the *essence* of this extraterrestrial world. I believe it requires a refocusing of the mind — which I have not yet achieved, but conceivably could, even while staying dry.

So I let the sun tattoo secret places of skin, feel the wind in the coconut palms wiping memory clean, hear the trilling voices of noddy terns working the shallows as meditation bells, and the rumble of the big surf on the oceanside a mantra. In theory, just as I absorb these, I could surrender, just as you might surrender to the blissful and silent presence of a guru. Then, perhaps, understanding would follow. Instead, because it's hot, I step from the sand on the beach into the sand under the water. The two worlds join almost seamlessly because there is virtually no difference between the air temperature and the water temperature, between the moisture in the air and that in the sea.

Teti'aroa's lagoon is fully enclosed within its barrier reef, with no big *hoa* from which large animals can easily come and go, and so this miniature atoll offers little excitement, not too much motion, only rarely any action that might invigorate the questing brain. For once, I don't even bother to don my facemask, preferring, for this moment, to see things as they are through my own landborne eyes. The water is thick with particulate, giving gentle, enclosed Teti'aroa the look of amniotic fluid.

Truly, there are many souls embodied in water, and here, among out-of-focus outcroppings of pink and purple baby corals, I find the blurry shapes of juvenile fish who surfed here some recent moonless night, even as new youngsters entered the next night, and the next, so that life begins here and then continually begins again. A school of juvenile surgeonfish resides upon a tiny lavender coral, a small school already gaudy in yellow and blue — perhaps the offspring of the spawners at Rangiroa Atoll two hundred miles north.

Afloat in this buoyant world, it occurs to me that life, which by its nature is forever evolving new forms, new strategies, new behaviors, new interdependencies, becomes (or else always has been) so complex that we will never catch up and fully understand it. Perhaps, in the end, this is the essence of the reef's rapture: the submission to the unknown, to a childlike sense of wondrous ignorance.

Sensing something in the water behind me, I turn and see a newborn blacktip reef shark, still soft and rubbery from the womb, hunting the bottom. In my maskless myopia, it appears to have a fourth dimension to it: that of its motion. As it turns to chase a shoal of baby convict tang toward me, then between my feet, past my shoulder, and around the top of my head, I can see — or seem to see — shadows of the pursued and the pursuer, like ghosts trailing a millisecond behind. Of course, it's only my own faulty eyes. But in the wake of their passing, I feel the currents of water ripple the length of my body, as if the energy of their struggle, of the lifeforce itself, has stroked invisible fingertips across me.

Notes

page Part I. Rangiroa

1   *"We feel surprise":* Charles Darwin in Leonard Engel, ed., *The Voyage of the Beagle* (Garden City, NY: Anchor Books, Doubleday and Company, Inc., 1962), 463–64.

5. BREATH CONTROL

30 *The client appears:* Redouan Bshary and Manuela Würth, "Cleaner Fish *Labroides dimidiatus* Manipulate Client Reef Fish by Providing Tactile Stimulation," *Proceedings of the Royal Society B* 268 (2001): 1495–1501.
    Clients are less likely: Redouan Bshary and A. S. Grutter, "Asymmetric Cheating Opportunities and Partner Control in a Cleaner Fish Mutualism," *Animal Behaviour* 63 (2002): 547–55.

31 *single cleaner fish servicing:* Alexandra S. Grutter, "Cleaner Fish Really Do Clean," *Nature* 398 (1999): 672–73.
    cleaners share an unusual color: Justin Marshall, personal communication.

35 *"Don't let your throat":* From an unnamed poem. Coleman Barks, trans., with John Moyne, *The Essential Rumi* (Edison, NJ: Castle Books, 1997), 12.

7. THE NEAR-FIELD/FAR-FIELD BOUNDARY

49 *roughly sixty earthquake-free days:* Junkee Rhie and Barbara Romanowicz, "Excitation of Earth's Incessant Free Oscillations

by Atmosphere-Ocean-Seafloor Coupling," *Nature* 431 (2004): 552–56.

8. EAVESDROPPING

57 *His work reveals:* "Christopher Clark: Whales of Newfoundland can hear whales near Bermuda," www.news.cornell.edu/chronicle/05/2.24.05/AAAS.Clark.whales.html.
tracking an unknown loner: William A. Watkins, M. A. Daher, J. E. George, and D. Rodriguez, "Twelve Years of Tracking 52-Hz Whale Calls from a Unique Source in the North Pacific," *Deep Sea Research I* 51 (2004): 1889–1901.

11. IMPERMANENCE

76 *"This day is already done":* This is from an examination of impermanence in *Buddhism A to Z,* compiled by Ronald B. Epstein (Burlingame, CA: Buddhist Text Translation Society, 2003), 115.

15. THE CONSORTING TOGETHER OF DISSIMILAR ORGANISMS

105 *"unifies the parts":* Lin Yutang, ed. and trans., *The Wisdom of Laotse* (New York: The Modern Library, Random House, Inc., 1948), 244–45.

106 *the presence of a cyanobacteria:* Michael P. Lesser, Charles H. Mazel, Maxim Y. Gorbunov, and Paul G. Falkowski, "Discovery of Symbiotic Nitrogen-Fixing Cyanobacteria in Corals," *Science* 305 (2004): 997–1000.

17. THE LEMON SHARK AFFAIR

120 *A study on Bahamian reefs:* Mark A. Hixon and M. H. Carr, "Synergistic Predation, Density Dependence, and Population Regulation in Marine Fish," *Science* 277 (1997): 946–49.

18. GRAND SECRET

122 *"an untriggered bomb":* Bengt Danielsson, "Poisoned Pacific: The Legacy of French Nuclear Testing," *Bulletin of the Atomic Scientists* 46 (1990): 22–31.

Part II. Funafuti

131 *"What blossoms":* This poem is by Ono No Komachi. Jane Hirshfield and Mariko Aratani, trans., *The Ink Dark Moon* (New York:
Vintage Books, Random House, Inc., 1990), 26.

19. HIDEAWAY

138 *"Although the wind":* This poem is by Izumi Shikibu. Ibid., 124.

20. FALLING DOMINOES

140 *"Let the hurricane tear":* Charles Darwin in Leonard Engel, ed.,
The Voyage of the Beagle (Garden City, NY: Anchor Books, Dou-
bleday and Company, Inc., 1962), 459.
"At Keeling atoll": Ibid., 473–74.

144 *"To what":* Peter Harris, ed., *Zen Poems* (New York: Everyman's Library Pocket Poets, 1999), 75.

21. LIQUID FAULTLINE

147 *"Come quickly":* This poem is by Izumi Shikibu. Hirshfield and
Aratani, trans., *The Ink Dark Moon,* 96.

22. LEAVE YOUR VALUES AT THE FRONT DESK

153 *"Yesterday":* This poem is by Izumi Shikibu. Ibid., 99.

24. DIVING THE APOCALYPSE

167 *some polyps repartner:* Andrew Baker, "Reefs Get Global Warming
Lifeline," *Nature* 430 (2004): 742.

27. BESEECHING THE WIND HORSES

187 *"Tuvaluans are blessed":* Excerpted from foreword to *Time and
Tide: The Islands of Tuvalu,* by Peter Bennets and Tony Wheeler
(Melbourne: Lonely Planet).

189 *"Tuvalu's voice in the debate":* Ibid.

28. SINKING DRAGONS

193 *planet's energy exchange system:* James Hansen, L. Nazarenko, R. Ruedy, M. Sato, J. Willis, A. Del Genio, D. Koch, A. Lacis, K. Lo, S. Menon, T. Novakov, J. Perlwitz, G. R. Russell, G. A. Schmidt, and N. Tausnev, "Earth's Energy Imbalance: Confirmation and Implications," *Science* 308 (2005): 1431–35.

195 *"Under the water":* This poem is by Izumi Shikibu. Hirshfield and Aratani, trans., *The Ink Dark Moon,* 76.

Part III. Mo'orea

199 *"In the ocean are many bright strands":* From a poem called "The Diver's Clothes Lying Empty." Barks, trans., with Moyne, *The Essential Rumi,* 51.

29. THE CHURNING OF THE OCEAN

203 *"It rises up from":* D'Arcy Wentworth Thompson, *Aristotle* (Oxford: Clarendon Press, 1910).

30. AN OCEAN OF SILENCE AND BLISS

211 *Compared to other dolphins:* This reference, and most others on the science of spinner dolphins, comes from a book by Kenneth S. Norris, Bernd Würsig, Randall S. Wells, and Melanie Würsig, *The Hawaiian Spinner Dolphin* (Berkeley: University of California Press, 1994).

213 *". . . every day ends the same":* Tenzin Wangyal Rinpoche, *The Tibetan Yogas of Dream and Sleep* (Ithaca, NY: Snow Lion Publications, 1998), 15.
"This state beyond duality": Alistair Shearer and Peter Russell, trans., *The Upanishads* (New York: Bell Tower, 2003), 38.

33. THE CLAMOR OF TRUE DEMOCRACY

227 *"it is better for the aware":* Tenzin Wangyal Rinpoche, *The Tibetan Yogas of Dream and Sleep,* 45.

34. THE SPIRIT OF GODLY GAMESOMENESS

231 *"There is something"*: Bernd Heinrich, *The Geese of Beaver Bog* (New York: Ecco, HarperCollins, 2004), i.

233 *"The observer"*: Kenneth S. Norris et al., *The Hawaiian Spinner Dolphin* (Berkeley: University of California Press, 1994), 91.

234 *"not merely a conventional system"*: Shearer and Russell, trans., *The Upanishads*, 15.

37. FISH TAMER

254 *larvae of coral reef fish:* Jeffrey M. Leis, "Vertical and Horizontal Distribution of Fish Larvae Near Coral Reefs at Lizard Island, Great Barrier Reef," *Marine Biology* 90 (1986): 505–16.

39. GLEANINGS

268 *the warming climate is triggering:* Catherine Drew Harvell, C. E. Mitchell, J. R. Ward, S. Altizer, A. P. Dobson, R. S. Ostfeld, and M. D. Samuel, "Climate Warming and Disease Risks for Terrestrial and Marine Biota," *Science* 296 (2002): 2158–62.

40. ACROSS THE THRESHOLD

273 *Doctoral research conducted in the 1990s:* Mary Gleason, "Coral recruitment patterns in Moʻorea, French Polynesia: The role of patch type and temporal variation," *Journal of Experimental Marine Biology and Ecology* 207 (1996): 79–101.

EPILOGUE

279 *"a slender pencil of land":* Quoted from Marlon Brando's autobiography, written with Robert Lindsey, *Brando: Songs My Mother Taught Me* (New York: Random House, 1994), 271.
"The happiest moments of my life": Ibid., 268.

Glossary

albedo: The fraction, measured between zero and one, of light or heat reflected by a surface.

algal ridge: The place on the barrier reef where the waves break; a zone too destructive for anything but encrusting red algae, or coralline algae, to grow.

Anekantavada: The Jain Doctrine of Manysidedness.

Aotearoa (Polynesian): Name for New Zealand.

arrowworm: A transparent predatory marine worm common in the zooplankton.

atoll: An island composed of a coral reef and low-lying sandy islands surrounding a lagoon.

autotroph: An organism that makes its food from light or chemical energy, transforming inorganic materials into organic life. All the primary producers (plants) are autotrophs, as are some marine invertebrates, including some corals, whose endosymbiotic zooxanthellae feed them.

barrier reef: A coral reef lying parallel to the shore and protecting both the shoreline and a lagoon.

bauplan (German): The archetypal body plan characterizing groups of animals.

benthic feeder: A marine animal that feeds on the bottom, usually by rooting through sand or mud.

bradycardia or **brachycardia:** Slowing of the heart rate by 50 percent or more.

caudal fin: The tail fin of fishes.

cellular respiration: The process whereby a cell converts organic compounds, usually sugars, into energy.

cephalopod: The group of mollusks including squids, octopuses, cuttlefishes, and nautiluses.

cetacean: A member of the order Cetacea; the whales, dolphins, and porpoises.

cilia: Hairlike structures.

clast: A fragment of a larger rock.

conspecifics: Members of the same species.

copepod: A tiny shrimplike crustacean.

coralline algae: Marine plants that produce calcium carbonate.

corallivore: An animal that eats only corals.

crustacean: An invertebrate of the class Crustacea, including the lobsters, crabs, shrimps, and barnacles.

ctenophore: A comb jelly or sea gooseberry; a jellyfishlike creature that generally wanders among the plankton, feeding on zooplankton. Most are the size of grapes and completely or mostly translucent.

Darwin Point: The point north of the Tropic of Cancer or south of the Tropic of Capricorn beyond which the water is generally too cold for corals to grow.

deep scattering layer (DSL): The vast community of life inhabiting the mesopelagic, from 650 to 3,200 feet deep, consisting of many types of fish, including lanternfish, lancetfish, krill, and squids of many kinds.

diurnal: Active during the daytime.

endosymbiosis: A symbiosis where one partner lives inside the cells of the other partner.

fale (Polynesian): A small house or hut.

falekaupule (Tuvaluan): A community hall.

fenua (Polynesian): Means "neighborhood," "island," "community," "country," as opposed to sea.

filter feeder: An animal that strains tiny food items from the water.

flash expansion: The breaking apart of a tightly packed shoal of fish as the individuals explode outward from an imaginary center; a defensive maneuver designed to confuse predators.

fringing reef: A coral reef growing close enough to the shore that no lagoon forms between it and the land.

gamete bundle: The hermaphroditic package of sperm and eggs produced by a spawning coral polyp.

guyot: The flat-topped remains of a submerged, extinct, oceanic volcano.

herbivore: An animal that eats only plants.

hoa (Polynesian): The shallow, watery channels separating *motu* in a coral atoll.

hypercarbia: High carbon dioxide levels in the blood.

Jainism: A religious philosophy that emerged from India in the sixth century B.C.; founded on an atheistic compilation of eternal truths.

kirikiri (Tuamotuan): The dark, tropical terns known as noddies.

laryngospasm: A protective reflex closing the larynx and preventing air or water from reaching the lungs.

lateral line: The line or lines along the sides of a fish containing sensory pores that detect vibrations.

lavalava (Tuvaluan): A sarong; the multipurpose sheet of cloth used by men and women as clothing.

lhun drub (Tibetan): "Spontaneous perfection."

live reef food fish trade (LRFFT): The trade in live coral reef fishes for sale in restaurants across Asia.

luciferin: The light-emitting pigments in bioluminescent organisms, including fireflies and marine animals of the deep sea.

Malthusian overfishing: A term for the extreme overfishing that occurs when the number of human fishers overwhelms the resource yet fishing continues in ever more destructive ways.

mammalian diving reflex (MDR): The response by mammals to sudden submersion in water; an autonomic energy-saving attempt to preserve life as the heart rate slows (bradycardia), as blood is rerouted from the extremities to the vital organs, inflating the chest.

mascaret (French): The high wave that travels backward up large rivers during flood tides; also refers to the sea currents within atoll passes that produce standing waves.

mesopelagic: The ocean between 650 and 3,200 feet deep; a zone of darkness with a constant temperature gradient regardless of season; sandwiched between the bathypelagic zone below and the epipelagic zone above.

motu (Polynesian): The low sandy island built on the flanks of a volcanic island by a living coral reef.

muro-ami fishery: A coral reef drive fishery originating in the Philippines; generally employing dozens or hundreds of children working in conditions of near slavery who free dive en masse to the reef, banging rocks and poles to drive the sea life into hanging nets.

myctophid: A bioluminescent lanternfish of the deep.

neritic zone: The ocean waters from the low-tide line to the edge of the continental shelf.

nyctinasty: The movement of a plant in response to darkness; usually involving the leaves folding or the petals closing.

oceanic zone: The oceanic waters lying beyond the continental shelves.

oligotrophic: Waters (or soils) that are poor in nutrients and have low primary productivity.

ota'a (Tuamotuan): Frigatebird.

paaihere (Tahitian): Fish known as jacks or trevallies, from the family Carangidae.

pêcheur (French): Fisherman.

pelagic zone: The open water away from sea coasts or the sea floor.

photophore: The light-producing organ found in some fishes (lanternfishes), cephalopods (squids, octopuses, cuttlefish), and crustaceans.

phytoplankton: The plant members of the plankton community.

piscivore: An animal that eats only or primarily fish.

planktivore: An organism that feeds primarily on plankton.

plankton: The drifting or weakly swimming animals and plants living at the mercy of currents.

planula: The tiny swimming larval stage of some corals and jellyfishes.

pranayama: The fourth limb of Patanjali's eightfold path of yoga, where breathing is consciously controlled, and where periods of breath-hold are interspersed into the rhythm of the inhalation and the exhalation.

pratyahara: The fifth limb of Patanjali's eightfold path, where consciousness is withdrawn from the senses, as a turtle withdraws its head, limbs, and tail into its shell.

primary producer: Any organism that uses inorganic (light or chemical) energy to make organic life; mostly refers to plants.

pteropod: A marine mollusk known as a sea butterfly or flapping snail.

pulmonary barotrauma: The lung damage caused to divers by ascending while holding the breath; a burst lung.

shadow zone: The stratum of water between the sun-warmed surface and the cooler water below where sound waves are refracted.

shield volcano: A gently sloping volcano built from fluid lava flows.

SOFAR: The sound fixing and ranging channel straddling the ocean's thermocline and the deep isothermic layer below it; a place where low-frequency sounds travel at a constant slow speed.

SOSUS or **sound surveillance system:** The U.S. Navy's underwater listening network composed of hydrophone arrays.

spur-and-groove zone: The place on the outer reef slope where the backwash from breaking waves wears deep canyons into the reef running perpendicular to the shore.

subsidence: The sinking or compression of land as a result of natural shifts (tectonic forces) or human activities (for example, water or oil extraction, bomb testing).

substrate: The surface where a plant or animal grows.

symbiosis: The close relationship, often obligatory, between two species, sometimes to mutual benefit (mutualism), but not always. Modern biology defines at least eight types of symbioses — parasitism, commensalism, aegism, epizoism, endoecism, phoresis, and inquilism — some less friendly than others.

teleosts: All the bony fishes from the subclass Teleostei.

terminal gasp: The fatal inhalation of a drowning victim after unconsciousness relaxes the laryngospasm and water floods the lungs.

thermocline: A boundary layer in a body of water where temperature decreases rapidly with depth.

thoracopod: The leg of a crustacean used for swimming, walking, feeding or filtering.

vaka or **va'a** (Polynesian): Outrigger canoe.

yugen (Japanese): the yearning for the sadness accompanying the loss of incomparably beautiful things.

zooxanthellae: The endosymbiotic algae living within the tissues of marine invertebrates, including corals; zooxanthellae provide the by-products of their photosynthesis (in the form of food energy or skeleton-building materials) to their animal partners, allowing some of these plant/animal hybrids to function as autotrophs, and awarding many their brilliant colors.